International Seminar on Strong and Electromagnetic Interactions in High Energy Collisions 2012

Journal of Physics: Conference Series Volume 424

Messina, Italy
12 October 2012

ISBN: 978-1-62748-301-8
ISSN: 1742-6588

Printed from e-media with permission by:

Curran Associates, Inc.
57 Morehouse Lane
Red Hook, NY 12571

Some format issues inherent in the e-media version may also appear in this print version.

Printed by Curran Associates, Inc. (2013)

Institute of Physics
Dirac House, Temple Back
Bristol BS1 6BE UK

Phone: 44 1 17 929 7481
Fax: 44 1 17 920 0979

techtracking@iop.org

Additional copies of this publication are available from:

Curran Associates, Inc.
57 Morehouse Lane
Red Hook, NY 12571 USA
Phone: 845-758-0400
Fax: 845-758-2634
Email: curran@proceedings.com
Web: www.proceedings.com

Preface

Papers

PREFACE

The International Seminar 'Strong and Electromagnetic Interaction in High Energy Collisions' was held in the Conference Hall 'Ettore Majorana' of the Department of Physics in Messina, Italy on October 12, 2012. The Seminar was organized by the University of Messina and 'Fondazione Bonino-Pulejo', with the aim of presenting and discussing the results of the current experiments and also new plans involving research at INFN-LNF (Italy), JLAB (USA), LHC-CERN, ELSA (Bonn), MAMI (Mainz).

The main purpose of this Seminar was to deal with aspects of electromagnetic and strong forces by meson photoproduction and the electron–positron collider, and to search for dark energy. The recent results on hadron contributions to the muon anomalous magnetic moment and kaon interferometry at the DAFNE facility were also discussed.

Editors:
Giorgio Giardina (University of Messina), Andrew M Sandorfi (Thomas Jefferson National Accelerator Facility, Newport News, USA), Paolo Pedroni (INFN 'Sezione di Pavia')

Organizing Committee:
Chairman: G Giardina (Messina - Italy)
Co-Chairman: A M Sandorfi (Newport News, USA)
Co-Chairman: P Pedroni (Pavia - Italy)

Scientific Secretary: G Mandaglio (University of Messina - Italy)

Organizing Institutions:
University of Messina
Fondazione Bonino-Pulejo (Messina)

Topics:
Meson photoproduction and baryon resonances
Muon anomaly (g-2)
Recent results in experiments at the Large Hadron Collider
Kaon interferometry

Local Organizing Committee:
F Curciarello, V De Leo, G Fazio, G Giardina, G Mandaglio, M Romaniuk

Sponsored by:
University of Messina, Fondazione Bonino-Pulejo (Messina), INFN Sezione di Catania

Web-Site: http://newcleo.unime.it/IntSem2012

Dates Issue 1 (2013)

Journal of Physics: Conference Series

Peer review statement

All papers published in this volume of *Journal of Physics: Conference Series* have been peer reviewed through processes administered by the proceedings Editors. Reviews were conducted by expert referees to the professional and scientific standards expected of a proceedings journal published by IOP Publishing.

Unravelling the Excitation Spectrum of the Nucleon

A.M. Sandorfi

Physics Division, Thomas Jefferson National Accelerator Facility, Newport News, VA 23606, USA

Email: sandorfi@JLab.org

Abstract. The low-energy structure of QCD lies encoded in the excited states of the nucleon, a complicated overlap of many resonances. Recent Lattice calculations have confirmed the long-standing quark model predictions of many more excited states than have been identified. Reactions that probe the spectrum are clouded by effects that *dress* the interactions and complicate the identification of excited levels and the interpretation of their structure. Recent theoretical work has exposed dramatic effects from such *dressings*. On the experimental side, new *complete* measurements of pseudoscalar meson photo-production are being pursued at several laboratories, where here the designation of *complete* refers to measurements of most if not all of the 16 possible reaction observables. This has been the focus of a series of experiments at Jefferson Lab culminating in the recently completed g9/FROST and g14/HDice runs which are now under analysis. With realistic errors, the number of observables needed to constrain the production amplitude is many more than required of a *mathematical* solution.

1. Quark Models and Lattice Calculations

A successful description of the excited levels of a composite system is a basic test of how well the underlying forces are understood. While Quantum-Chromodynamics (QCD) is generally regarded as a relatively mature theory describing the interactions of quarks within hadrons, the excited states of the nucleon pose many challenges. This partly arises because of the ways in which the spectrum, as well as the decay widths of excited levels, are influenced by the underlying nature of quark confinement and Chiral Symmetry breaking, and partly because of multiple effects that *dress* the interactions and alter their manifestation. This paper discusses some of the features and puzzles associated with excited nucleons and describes a new experimental initiative aimed at unfolding the nucleon spectrum through measurements of *everything*.

Quark models have been very successful at capturing the symmetries of QCD without solving its Lagrangian. In such models, a nucleon has 3 *constituent* quarks with masses of about 0.3 GeV. At the same time the *current* quarks of deep-inelastic scattering have energies of at most a few MeV. Recent theoretical calculations comparing QCD on a Lattice with solutions of the Dyson-Schwinger equations (DSE) have provided the connection between these two successful but very different pictures [1]. Figure 1 shows the running quark mass, dressed by gluons, as a function of momentum. The descending sets of points are Lattice-QCD (LQCD) calculations that start with different current quark masses in the perturbative regime, as indicated in the legend to the right of the plot, the lowest being a linear extrapolation to the Chiral limit of zero. The curves are the corresponding DSE results. The accumulation of effective mass below about 1 GeV/c arises from a cloud of gluons that are dragged along with the quark as it propagates. The IR inflection that is evident in the DSE curves below about ¼ GeV/c results from Dynamical Chiral symmetry breaking and is the signature of quark confinement.

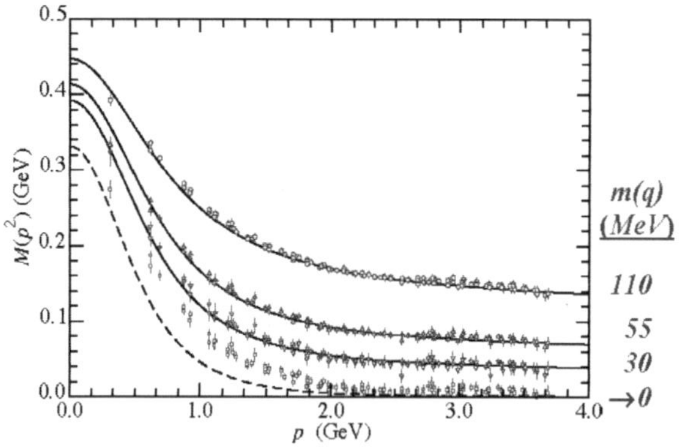

Figure 1. Effective quark mass as a function of momentum, from [1]. Descending sets of Lattice QCD points correspond to calculations with different assumptions for the current quark mass, as indicated to the right. The solid curves are the corresponding Dyson-Schwinger calculations. The *constituent* mass appears at low momentum from the gluon cloud.

The results of figure 1 have provided the anticipated connection between the constituent mass of *QCD-inspired* models, which on the whole are quite successful, and the current quarks of QCD. Specifically, it is the gluon cloud, emerging from the dynamical breaking of Chiral symmetry, that generates *constituent* quarks with effective masses and the essential features captured in quark models.

A long-standing prediction of constituent quark models (CQM) is the expectation of many more N* levels than have been observed. The predicted spectrum from Capstick and Roberts [2] is reproduced in the left panel of figure 2. Candidate πN resonances assigned a 3 or 4-star status by the Particle Data Group are circled. Only the lowest few states in each oscillator band correspond to observed resonances.

A traditional model approach to the apparent paucity of identified N* resonances has been to invoke mechanisms that effectively reduce the number of degrees of freedom. Diquark correlations, in which two quarks are coupled to a *point-like*, spin 0 and isospin 0 object, have been employed for some decades. By freezing out internal diquark excitations the number of predicted levels are reduced significantly [3]. However, more recent calculations with the Dyson-Schwinger equations have shown that diquark correlations, which are indeed energetically favored and hence very likely, have about the radius of a pion [4]. Thus internal diquark excitations are not expected to be frozen out.

Recent LQCD predictions from the *Hadron Spectrum Collaboration* [5] are shown for comparison in the right panel of figure 2. These LQCD calculations include all possible *q-q* correlations and still yield a large number of as yet to be discovered N* states. In short, *diquarks can no longer be regarded as a solution to missing N* states.* With this LQCD validation of predictions for a plethora of higher

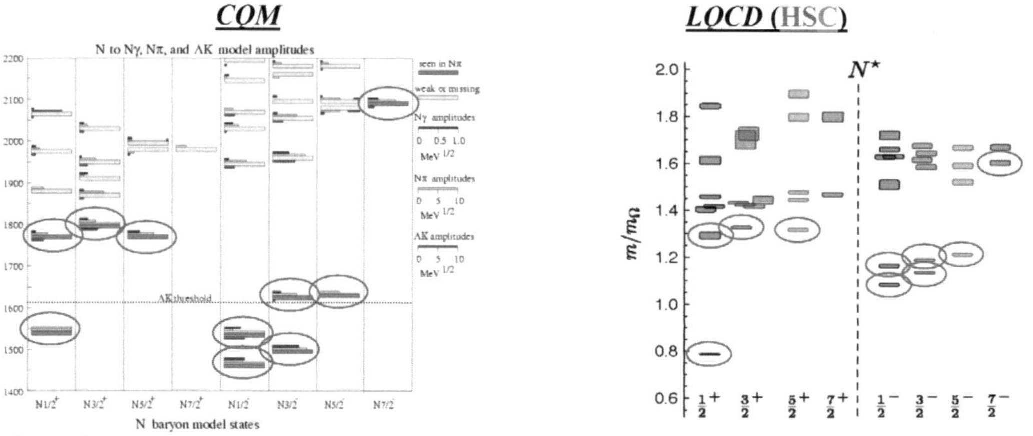

Figure 2. The N* spectrum as predicted by the CQM (left panel), adapted from [2], compared with LQCD predictions (right panel) from [5]. Candidate πN resonances with 3 or 4-star status are circled.

Journal of Physics: Conference Series **424** (2013) 012001 doi:10.1088/1742-6596/424/1/012001

lying levels, uncovering new resonances has become a high priority. Indeed, their absence at this point would pose serious problems.

2. Nature's *Dressings*

There are a number of factors that can complicate the search for new N* levels. First of all, the CQM predicts (and by inference, the symmetry of the QCD Lagrangian expects) πN partial widths that diminish with energy. This may account for the difficulty in isolating signatures of higher lying N* states in πN scattering data. Fortunately, these same higher lying states are expected to have larger branches to KY and ππN final states. Such final states are excited in photo-production and the spin of the photon provides additional probes of the amplitude.

Further complications in comparing photo-production processes to a predicted spectrum enter through what are usually called *dressing* of the interaction vertices. We associate a *resonance* with a pole in an *s*-channel diagram, such as in figure 3. But a lot of detail is buried in the vertices. Models of

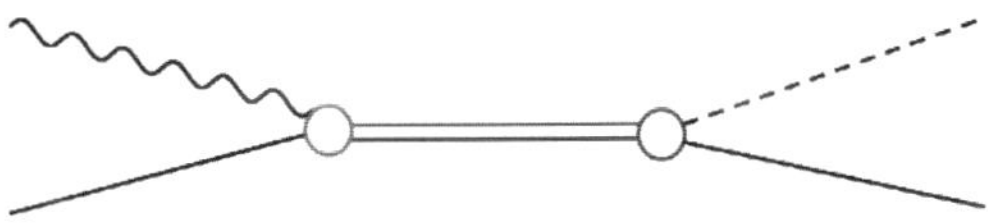

Figure 3. An ***s*-channel pole** excited through an **electromagnetic vertex**, with subsequent decay through the **strong vertex**.

the electromagnetic vertex must incorporate effects from the dynamical breaking of chiral symmetry which dresses the neutron and proton with a cloud of spin zero mesons (π, η, κ). For example, the photon of figure 3 can attach to a nucleon that is off-shell due to the presence of a virtual pion loop, or it can attach to the pion itself, etc. So the initial vertex is really an expansion. At the present level of sophistication, quark models do not include such pion clouds, so differences from measurement will be expected. In general, such mesonic *dressings* of the electromagnetic vertex affect the dynamical properties of a resonance and the Q^2 evolution of N* production, but do not affect spectral properties.

The higher energies of the *missing* N* levels are above thresholds for many decay channels. Hyperon channels, for example, can be either produced directly (*eg*. $\gamma p \to K\Lambda$) or reached through various two-step processes (*eg*. $\gamma p \to \pi N \to K\Lambda$). As a result, calculating the *strong* vertex of figure 3 becomes a multi-channel, multi-resonance problem, with Unitarity requirements connecting all possible channels. These *coupled-channel dressings* of the strong vertex essentially determine the spectral properties of N* resonances – their mass/pole positions and widths.

We can illustrate the challenges and the fruits of detailed reaction analyses using the example of the $P_{11}(1440)$ Roper resonance, the nucleon's second excited level. This state has a complicated history and has long been shrouded in mystery. It was one of the first examples of a resonance that is not directly observable as a peak in a cross section, but rather required by detailed partial wave analyses (PWA) [6]. Recent PWA in fact suggest a doublet of closely spaced poles [7]. As a general rule, all calculations predict this state at too high an energy. It has been modeled mostly as either a 3-quark radial excitation (a nucleon *breathing mode*) or as a *qqq-g* quark-gluon hybrid. These dynamical distinctions can be separated by studying the Q^2 evolution of the electro-production amplitude. The P_{11} decays into πN, ππN and ΔN final states. The helicity amplitude for P_{11} electro-production is shown in figure 4 as a function of the square of the 4-momentum transfer (Q^2). Data from the CLAS detector at Jefferson Lab is plotted for both the πN (red circle) [8] and the ππN (blue square) [9] final states. The green short-dashed curve gives the expected Q^2 dependence of a *qqq-g* gluonic-hybrid excitation from [10]. This picture completely misses the cross-over in sign at around ½ GeV^2 and can be ruled out. The shaded band shows the range of various quark-model calculations, adapted from [9]. These exhibit the change in sign at low Q^2 (a consequence of relativity) and the falling trend at high Q^2, but appear shifted from the data. However, these quark models do not contain pion clouds. The expected contributions from the nucleon's virtual pion cloud are shown as the orange long-dashed curve [11] in figure 4 and are quite substantial. Combined with the appropriate phase, they can at least qualitatively

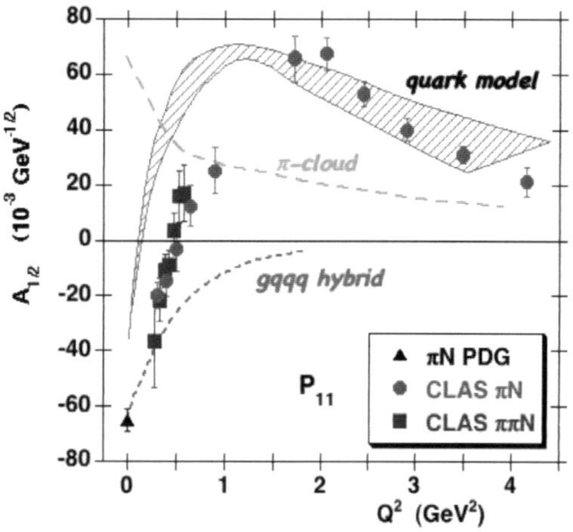

Figure 4. Helicity amplitudes for $P_{11}(1440)$ electro-production, deduced from πN (red circles) [8] and ππN (blue squares) [9] final states. The short-dashed green curve gives the *qqq-g* quark-gluon hybrid expectation from [10]. The shaded band shows the range of various quark-model calculations, adapted from [9] and references therein. The expected contribution from the pion cloud is indicated as the orange long-dashed curve [11].

account for the shift in quark-model calculations that do not include meson clouds. In this way, Q^2 evolutions probe the photon vertex of figure 3 and provide valuable information on excitation mechanisms.

The meson-baryon dressings of the strong vertex of figure 3 can only be studied in combined simultaneous *coupled-channel* analyses that maintain Unitarity across many meson decay channels. Recent calculations from the group at the Excited Baryon Analysis Center (EBAC) at Jefferson Lab have used a dynamic coupled channel model to simultaneously describe γN and πN reactions leading to πN, πΔ, ρN, σN, ηN, KΛ and KΣ final states. This calculation starts with a single *bare* N* state (equivalent to a quark model level) in the P_{11} partial wave and analytically continues this into the complex plane. (Here, a grid search is carried out to find minima in the denominator of the *T* matrix as the strengths of the channel couplings are allowed to grow. In this way, trajectories in the complex plane are mapped out leading to poles on multiple Riemann sheets.) Their results for the poles in the final state P_{11} amplitude are shown in figure 5 [12]. As meson-baryon channel couplings are increased, what starts off as a bare quark excitation at 1763 MeV splits into a high-lying P_{11} state and a closely spaced doublet at the energy of the *Roper*, one dominated by the πN channel and the other by ππN. This doublet nature of the *Roper* resonance is in fact a common feature of all recent multipole analyses. In addition, it is now apparent why all previous N* models had difficulty accounting for the low energy of the *Roper* resonance – the excitation of the quark core has to occur at a considerably higher energy in order to give the channel couplings the space to bring the apparent pole position(s) down to the energy of the experimentally observed resonance(s). In recent DSE calculations, this *bare* P_{11} is in fact generated by a radial excitation of a diquark, coupled to the third quark of the nucleon.

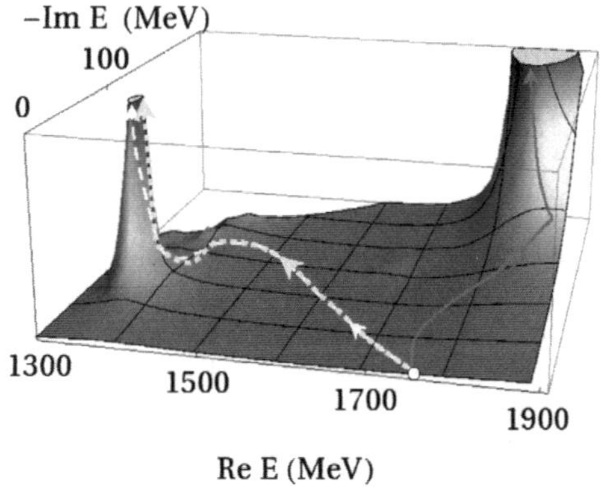

Figure 5. The trajectories and poles in the P_{11} amplitude, as calculated in the EBAC Dynamic Coupled-Channel model of meson production. (Adapted from [12].) Channel-couplings cause the *bare* N* state at 1763 MeV to split into a high lying P_{11} and a doublet corresponding to the traditional *Roper* resonance.

3. Measurements of *Everything*

Single pseudo-scalar photoproduction is described by 4 complex amplitudes. A mathematical solution for the amplitude would require angular distribution measurements of at least 8 carefully chosen observables at each energy for both proton and neutron targets [13]. The lack of such experimental information, despite 50 years of photo-production measurements, has created ambiguities that could easily account for the difficulties in locating N* resonances. To address this shortcoming, a sequence of *complete* experiments have been completed and are now under analysis at Jefferson Lab [14,15], as well as complementary experiments at Bonn and Mainz, with the goal of obtaining a direct determination of the amplitude to within a phase, for at least a few production channels.

In single-pseudoscalar meson photoproduction there are 16 possible observables, the cross section (σ), three asymmetries which to leading order enter the general cross section scaled by a single polarization of either beam, target or recoil (Σ, T, P), and three sets of four asymmetries whose leading dependence in the general cross section involves two polarizations of either beam-target (E, G, F, H), beam-recoil ($C_{x'}$ $C_{z'}$, $O_{x'}$, $O_{z'}$) or target-recoil ($L_{x'}$, $L_{z'}$, $T_{x'}$, $T_{z'}$). However, each of these observables enters the general form of the cross section scaled by two terms containing different combinations of beam, target and recoil spins [16]. This leads to two distinct ways to measure each observable. The full set are summarized in Table 1. There, a caret (^) is used to designate the numerator of an asymmetry, $\hat{A} = A \cdot \sigma_0$. The bottom four rows correspond to different states of beam polarization, either without regard to incident polarization (*unpol*) or in one of the three standard *Stokes* conditions that characterize the possible polarizations of an ensemble of photons, linear with a $sin(2\phi_\gamma)$ dependence relative to the reaction plane, linear with a $cos(2\phi_\gamma)$ dependence, or circular. The columns of the table give the polarization of the target, recoil, or target+recoil combination. The signs indicate the sign of the corresponding term entering the general cross section. One can readily construct from this table the terms that enter the general cross section for any given combination of polarization conditions. For example, with linear beam polarization in or perpendicular to the reaction plane, a longitudinally polarized target (along z) and an analysis of recoil polarization along the meson momentum (z'), the general cross section is given by terms in the first (*unpol*) and third rows that are either independent of target and recoil polarization (σ_0, Σ) or in columns associated with polarization along z and/or z',

Table 1. Polarization observables in pseudoscalar meson photo-production. Each appears twice here. Carets (^) designate the products of asymmetries with the cross section, $\hat{A} = A \cdot \sigma_0$. The 16 in black indicate the leading polarization dependence of each observable to the cross section. The three in red (P, T, Σ) are nominal *single-polarization* quantities that can be measured with double-polarization. Those in blue are the unpolarized cross section and 12 nominal *double-polarization* quantities that can be measured with triple-polarization. Signs indicate the sign of each term as it enters the general cross section (from [16]).

		Target (P^T)			***Recoil*** (P^R)			***Target*** (P^T) + ***Recoil*** (P^R)								
					x'	y'	z'	x'	x'	x'	y'	y'	y'	z'	z'	z'
γ ***beam*** (P^γ)		x	y	z				x	y	z	x	y	z	x	y	z
unpol :	σ_0		$\hat{T}$			$\hat{P}$		$\hat{T}_{x'}$		$\hat{L}_{x'}$		$\hat{\Sigma}$		$\hat{T}_{z'}$		$\hat{L}_{z'}$
P_L^γ ***sin*** $2\phi_\gamma$:		$\hat{H}$		$\hat{G}$	$\hat{O}_{x'}$		$\hat{O}_{z'}$		$\hat{C}_{z'}$		$\hat{E}$		$\hat{F}$		$-\hat{C}_{x'}$	
P_L^γ ***cos*** $2\phi_\gamma$:	$-\hat{\Sigma}$		$-\hat{P}$			$-\hat{T}$		$-\hat{L}_{z'}$		$\hat{T}_{z'}$		$-\sigma_0$		$\hat{L}_{x'}$		$-\hat{T}_{x'}$
cir P_c^γ :		$\hat{F}$		$-\hat{E}$	$\hat{C}_{x'}$		$\hat{C}_{z'}$		$-\hat{O}_{z'}$		$\hat{G}$		$-\hat{H}$		$\hat{O}_{x'}$	

$$d\sigma(P^{\gamma},P^{T},P^{R})=\tfrac{1}{2}\left\{\left[\sigma_0+P_z^T P_{z'}^R \hat{L}_{z'}\right]+P_L^{\gamma}\cos(2\phi_\gamma)\left[-\hat{\Sigma}-P_z^T P_{z'}^R \hat{T}_{x'}\right]\right\}.$$

The initial factor of ½ is always present and ensures that the unpolarized cross section is recovered when averaging over initial states.

There is some confusion in the literature regarding definitions of asymmetries and the conventions used in partial wave analyses (PWA). This arises from the use of different coordinate systems and because the same symbol or *asymmetry name* has been used by different authors to refer to different experimental quantities. Expressions for the photo-production asymmetries as coordinate-system independent ratios of cross sections, along with the names used for these ratios by different PWA groups, can be found in ref [17]. Examples of the explicit measurements needed to construct all of the observables are given in the appendices to ref [16].

4. The special role of Recoil Polarization

In pseudoscalar meson photo-production, the beam, target and recoil baryon can all be polarized, although these polarizations are not all on an identical footing. While the beam and target polarization are under experimental control, the recoil polarization is rather a by-product of the entrance channel angular momentum and the reaction physics. Expressions for the recoil polarization can be derived by grouping together the terms in the general cross section that dependent upon the components of the recoil polarization vector [16],

$$d\sigma^{\mathrm{B,T,R}}(\vec{P}^{\gamma},\vec{P}^{T},\vec{P}^{R})=\tfrac{1}{2}\left[A^0+(P_{x'}^R)A^{x'}+(P_{y'}^R)A^{y'}+(P_{z'}^R)A^{z'}\right],$$

where,

$$\begin{aligned}
A^0 &= d\sigma_0 - P_L^{\gamma}\cos(2\phi_\gamma)\hat{\Sigma} + P_y^T\hat{T} - P_L^{\gamma}P_y^T\cos(2\phi_\gamma)\hat{P} - P_c^{\gamma}P_z^T\hat{E} + P_L^{\gamma}P_z^T\sin(2\phi_\gamma)\hat{G}\\
&\quad + P_c^{\gamma}P_x^T\hat{F} + P_L^{\gamma}P_x^T\sin(2\phi_\gamma)\hat{H},\\
A^{x'} &= P_c^{\gamma}\hat{C}_{x'} + P_L^{\gamma}\sin(2\phi_\gamma)\hat{O}_{x'} + P_z^T\hat{L}_{x'} + P_x^T\hat{T}_{x'} + P_L^{\gamma}P_y^T\sin(2\phi_\gamma)\hat{C}_{z'} - P_c^{\gamma}P_y^T\hat{O}_{z'}\\
&\quad - P_L^{\gamma}P_x^T\cos(2\phi_\gamma)\hat{L}_{z'} + P_L^{\gamma}P_z^T\cos(2\phi_\gamma)\hat{T}_{z'},\\
A^{y'} &= \hat{P} + P_y^T\hat{\Sigma} - P_L^{\gamma}\cos(2\phi_\gamma)\hat{T} - P_L^{\gamma}P_y^T\cos(2\phi_\gamma)d\sigma_0 + P_L^{\gamma}P_x^T\sin(2\phi_\gamma)\hat{E} + P_c^{\gamma}P_x^T\hat{G}\\
&\quad + P_L^{\gamma}P_z^T\sin(2\phi_\gamma)\hat{F} - P_c^{\gamma}P_z^T\hat{H},\\
A^{z'} &= P_c^{\gamma}\hat{C}_{z'} + P_L^{\gamma}\sin(2\phi_\gamma)\hat{O}_{z'} + P_z^T\hat{L}_{z'} + P_x^T\hat{T}_{z'} - P_L^{\gamma}P_y^T\sin(2\phi_\gamma)\hat{C}_{x'} + P_c^{\gamma}P_y^T\hat{O}_{x'}\\
&\quad + P_L^{\gamma}P_x^T\cos(2\phi_\gamma)\hat{L}_{x'} - P_L^{\gamma}P_z^T\cos(2\phi_\gamma)\hat{T}_{x'}.
\end{aligned}$$

With this arrangement it is straight forward to show that the components of the recoil polarization vector are given by [16],

$$P_{x'}^R=\frac{A^{x'}}{A^0},\qquad P_{y'}^R=\frac{A^{y'}}{A^0},\qquad P_{z'}^R=\frac{A^{z'}}{A^0}.$$

This can lead to a particularly fruitful arena in which different components of the recoiling baryon's polarization determine various spin observables. The two most practical reactions for this are in KY production, where the angular distribution in the weak decay of the hyperon provides a self-analysis of the polarization,

$$\gamma\,\mathrm{N} \rightarrow \mathrm{K}\,\Lambda \Rightarrow \mathrm{K}\,(\pi^{-}\mathrm{p})\,,$$
$$\gamma\,\mathrm{N} \rightarrow \mathrm{K}\,\Sigma^{+} \Rightarrow \mathrm{K}\,(\pi^{0}\mathrm{p})\,.$$

The K Λ channel has an analyzing power of about 2/3 while that of the $K\pi^0p$ decay branch in the K Σ^+ channel is almost unity. When a quasi-4π detector is used, such as the CLAS at Jefferson Lab, recoil polarization information is in the data stream and comes without a significant penalty in statistics.

It is straight forward to use the above expressions to deduce observables involving recoil polarization. The simplest case occurs when neither the beam nor the target are polarized. Then the *y* component of the recoil polarization nonetheless survives and is determined by the observable *P*. This is a recoil polarization generated solely by the reaction physics and can be non-zero. An example from the $\gamma p \rightarrow K^+\Lambda \rightarrow K^+\pi^-p$ reaction is shown schematically in figure 6. In the rest frame of the Λ, the angular distribution of its weak decay into π^-p takes the form,

$$W(\theta_p) = \tfrac{1}{2}\left(1+\alpha\left|\vec{P}_\Lambda\right|\cos(\theta_p)\right).$$

Here θ_p is the angle between the proton direction and the normal to the reaction plane in the rest frame of the Λ and α is the Λ decay parameter, α = 0.642 ±0.013. Since the cosine function changes sign between quadrants, a measurement of the recoil polarization consists in simply counting protons above and below the reaction plane.

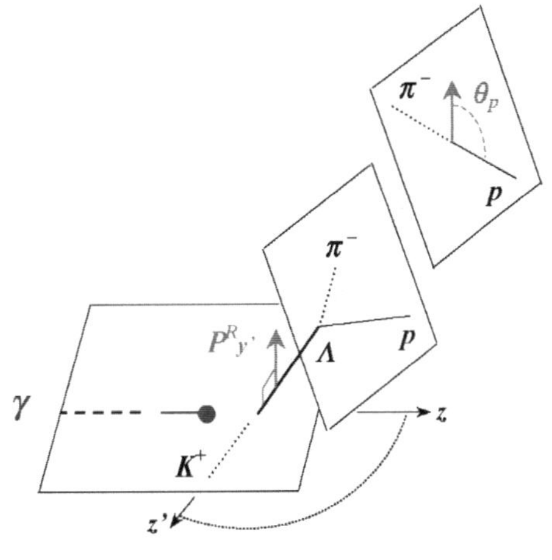

Figure 6. With beam and the target unpolarized the recoil polarization can only be perpendicular to the reaction plane, $\vec{P} = (0, P, 0)$ and is determined by the proton yields on either side of the reaction plane,

$$P = \frac{2}{\alpha}\frac{N_{up} - N_{dwn}}{N_{up} + N_{dwn}}\,.$$

As an example of this technique, recent measurements of the recoil asymmetry *P* are plotted in figure 7 for W=1728 MeV in the $\gamma p \rightarrow K^+\Lambda$ reaction. The data were collected in the *BGO-Ball* at GRAAL [18] and with the CLAS detector at Jefferson Lab [19]. As mentioned above, with quasi-4π detectors the weak decay information is in the data stream and quite accurate results can readily be obtained.

Several other examples with different experimental conditions are worked out in ref [16]. The most fruitful conditions arise when both the beam and the target are polarized. Then, the recoil polarization contains a wealth of information. For example, when circularly polarized photons are used with a longitudinally polarized target (with target spin along the beam direction), the above equations determine the components of the recoil polarization as,

$$\vec{P}^R = \left(\frac{P_c^\gamma C_{x'} + P_z^T L_{x'}}{1 - P_c^\gamma P_z^T E},\ \frac{P - P_c^\gamma P_z^T H}{1 - P_c^\gamma P_z^T E},\ \frac{P_c^\gamma C_{z'} + P_z^T L_{z'}}{1 - P_c^\gamma P_z^T E}\right).$$

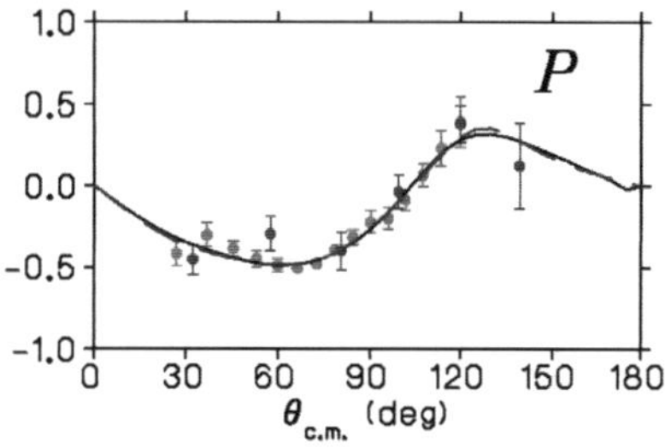

Figure 7. The recoil polarization asymmetry in the $\gamma p \to K^+\Lambda$ reaction at W=1728 MeV. Blue data points are GRAAL measurements from [18]; red data are CLAS results from [19].

The asymmetry in the denominator (E in red above) and the 6 observables in the numerator (in blue) can be extracted from successive analyses of the data set by choosing subsets with different spin orientations. When the recoil polarization is ignored and final states are summed, yields with the beam and target spins anti-parallel and parallel determine the E asymmetry and the denominator in the above relations. Combining flux-weighted subsets of data with opposite target polarizations, $\pm P_z^T$, determines $C_{x'}$, P and $C_{z'}$. Another pass through the data to average initial beam polarization states, $P_c^\gamma(h=\pm1)$, then determines $L_{x'}$ and $L_{z'}$. Finally, the difference between flux-weighted subsets with beam and target spins parallel and anitparallel, $P_c^\gamma(P_{+z}^T - P_{-z}^T)$, determines the H asymmetry. Interestingly, H is traditionally discussed as a *Transverse Target* asymmetry, since it appears in the second observable column of Table 1. Here, it is constructed from measurements with a longitudinally polarized target, utilizing its appearance in the forth-last column. In a similar manner, utilizing also linearly polarized beams, all 16 observables can be measured with a single target orientation [16].

5. The FROST and HDice experiments at Jefferson Lab

The key to determining complete sets of observables are triple polarization measurements. Circularly polarized photon beams are routinely created by the bremsstrahlung of longitudinally polarized electrons and linear beam polarization is generated through coherent bremsstrahlung in single crystals. As we have seen above, the analysis of weak decays in quasi-4π detectors determines recoil polarization. The most challenging aspect of such experiments is the polarized target.

At Jefferson Lab, the *g9* experimental runs focused on proton reactions and utilized a frozen-spin target ("FROST") of butanol, C_4H_9OH. Here, the electrons of butanol were polarized at 0.4K and 5 tesla, after which their polarization was transferred to the ten free H atoms of the molecule using a microwave transition. A subsequent drop in temperature to the 30 mK range produced a target with a relaxation time of about 100 days with a holding field of 0.5 tesla. This process was repeated about once per week to keep the butanol polarization high. This approximate *frozen-spin* state was used in two sets of experiments, with the H polarization held either along or transverse to the beam, respectively. Typical proton polarizations were at least 80% for the 10 free H nuclei in the mass 74 butanol molecule. Carbon foils were used to estimate the background contributions from the unpolarizable materials. Both g9 runs can be expected to independently yield a complete set of asymmetries for the $\gamma p \to K^+\Lambda$ channel.

Since the electromagnetic interaction does not conserve isospin, measurements with both proton and neutron targets are needed to completely specify the amplitude. For polarized neutron experiments, the recently completed JLab *g14* run used a target of molecular deuterium-hydride ("HDice") in the solid phase. These targets have many attractive features but are considerably more complex. The HD is initially loaded with small impurities (at the ~ 10^{-3} level) of H_2 and D_2. The first rotational states of these impurities are initially populated and metastable and polarize rapidly at the 0.010K and 15 tesla conditions of a large dilution refrigerator. The H_2 and D_2 spins exchange with spins of H and D in HD and then repolarize until all of the HD reaches high polarization. With time (~ a few months) the populations of the H_2 and D_2 rotational states decay to their magnetically inert ground states. At that point, the HD is in a true *frozen spin* state and can be transferred to an in-beam cryostat for experiments where relaxation times are essentially unmeasurable (~years) at temperatures below 0.1K and fields of 0.8 tesla. Initial H and D polarizations were typically 60% for the H and 15% for the D. However, spins can then be transferred between H and D with RF transitions. Since the g14 run focused on polarized neutron reactions, H $\Rightarrow$ D spin transfer was used and D polarizations up to 27% were achieved. Background reactions from the unpolarizable material of the target cell were small and could be directly measured by warming the target cell and pumping out the HD gas. The product of polarization and dilution (of unpolarizable material) in the HDice and FROST targets were comparable. Nonetheless, an order of magnitude higher rates could be used with HDice before reaching data acquisition limits, due to the reduced background of atomic electrons associated with the lower Z of this target. A single target orientation was used during the g14 run (along the beam direction) and is expected to yield a complete set of asymmetries for the $\gamma n \rightarrow K^0\Lambda$ channel.

Results for $\gamma\, p \rightarrow K^+\Lambda$ from the *g1c* and *g11a* running periods at Jefferson Lab using the CLAS with liquid hydrogen have already provided the cross section, P, $C_{x'}$ and $C_{z'}$ [19, 20] data. The inclusion of these in the Bonn-Gatchina PWA has lead to possible candidates for *new* states [21], although the solutions are still ambiguous. The CLAS *g8* period ran with linear photon polarization from coherent bremsstrahlung to measure Σ, T, P, $O_{x'}$ and $O_{z'}$ and these results are nearly finalized. Once the analyses of these measurements and the g9 and g14 results are complete, the results are expected to have dramatic impacts on deduced multipoles and on our understanding of the N* spectrum.

6. Amplitude extraction and uniqueness

Each of the 16 spin observables can be written in terms of 4 complex amplitudes, which can be expressed in ether Cartesian (F_i) [22], Spherical or Helicity (H_i), or Transversity (b_i) representations [23]. An explicit decomposition of all 16 into the 4 CGLN F_i amplitudes is given in [16]. However, the four amplitudes in each of these representations are angle dependent. Extracting them directly from experiment would require separate fits at each angle, which greatly limits the data that can be used and requires some model-dependent scheme to constrain an arbitrary phase that could be angle-dependent. The solution to this intractable situation is a *Wigner-Echart* style factorization into reduced matrix elements, or *multipoles*, and simple angle-dependent coefficients from angular momentum algebra [16,22]. One can then fit the multipoles directly, which both facilitates the search for resonance behavior and allows the use of full angular distribution data at a fixed energy to constrain angle-independent quantities. The price is a significant increase in the number of fitting parameters, but since the excited states of the nucleon are associated with discrete values for their angular momentum, this expansion of variables is inevitable.

Since the full amplitude is a sum of four independent complex terms, measurements of eight different observables are needed for a mathematical solution. (In practice, given realistic uncertainties on data, many more are required. But more on that later.) Since there are a total of 16 spin observables, it is clear that they are not all independent. A set of 37 *Fierz Identities* interrelate the spin asymmetries [13,16]. With a full suite of polarization data, the complete set of Fierz identities can be used to almost eliminate systematic scale shifts between data sets. However, utilizing them requires data sets at common angles and energies, which in practice requires some interpolation.

A critical issue for the goal of determining an amplitude is the uniqueness of the solution. The reaction channel with the largest number of different observables published to date is $\gamma\, p \rightarrow K^+\Lambda$, where results from GLAS *g1c* and *g11* [20,19] and from GRAAL [18,24] can be combined to provide values for eight observables $\{\sigma, \Sigma, T, P, O_{x'}, O_{z'}, C_{x'}$ and $C_{z'}\}$ over at least a limited W range, 1676 to 1883 MeV. We have carried out a multipole analysis for this combined set using a Monte Carlo sampling of the parameter space, coupled with gradient minimization [16]. These CLAS and GRAAL results determine bands of solutions, that are indistinguishable in their ability to reproduce the data but nonetheless derive from very different multipole amplitudes. As an example, data at W = 1883 MeV are plotted in figure 8 and compared to multipole solutions having the minimum (0.74) and largest

(0.92) χ^2/point. The corresponding solution bands are shown for multipoles up to L=2 in figure 9. The bands represent a shallow χ^2 valley pitted with a large number of local minima that are experimentally degenerate [16]. Evidently much more information will be required for a well-defined solution.

Amplitudes can of course be rotated by an arbitrary phase. There are two methods to remove such ambiguities. The simplest is to set the phase of one particular multipole to zero, and it helps to choose a relatively large one (*eg*. the E_{0+}, as has been done in the fits of figures 8 and 9). This does not generally limit the utility of a solution. To compare with results from other PWA one simply rotates all solutions

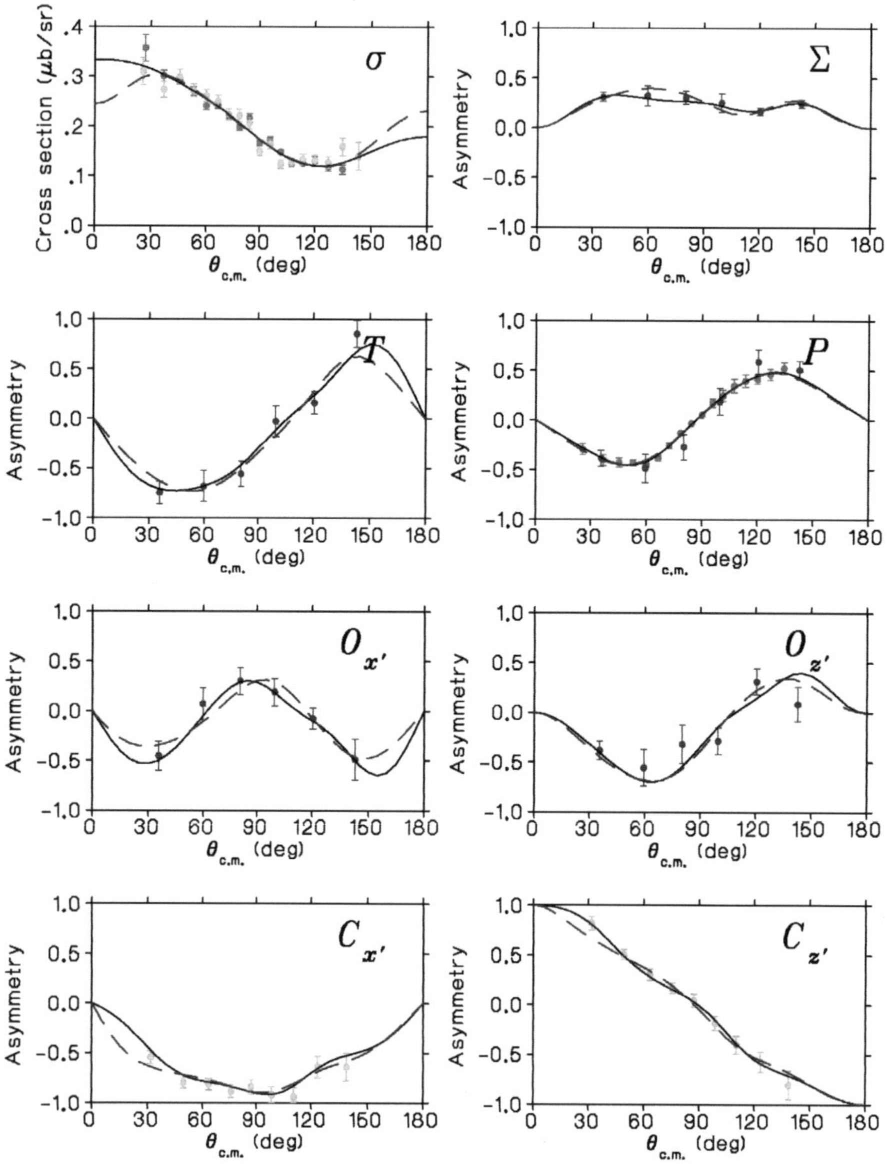

Figure 8. $\gamma p \rightarrow K^+\Lambda$ results from [20], [19], [18] and [24] at W=1883 MeV. The solid (black) and long-dashed (blue) curves are multipole fits with the lowest (0.74) and largest (0.92) χ^2/point.

to the same reference phase. (*eg*. Multiply all multipoles of another solution by $e^{-i\phi}$, where ϕ is the phase of the E_{0+} in that solution.) Another method of fixing the overall phase that has been successfully used in charged π production is to set high L multipoles to their real Born and *t*-channel pole values [25]. The *t*-channel graphs for π photo-production have significant well-established contributions that extent to very high *L* and these essentially determine the phase. Unfortunately, in K^+ production this procedure isn't directly applicable for several reasons: the *t*-channel couplings are much smaller (being inversely proportional to the mass of the exchanged meson), there are more meson candidates that can

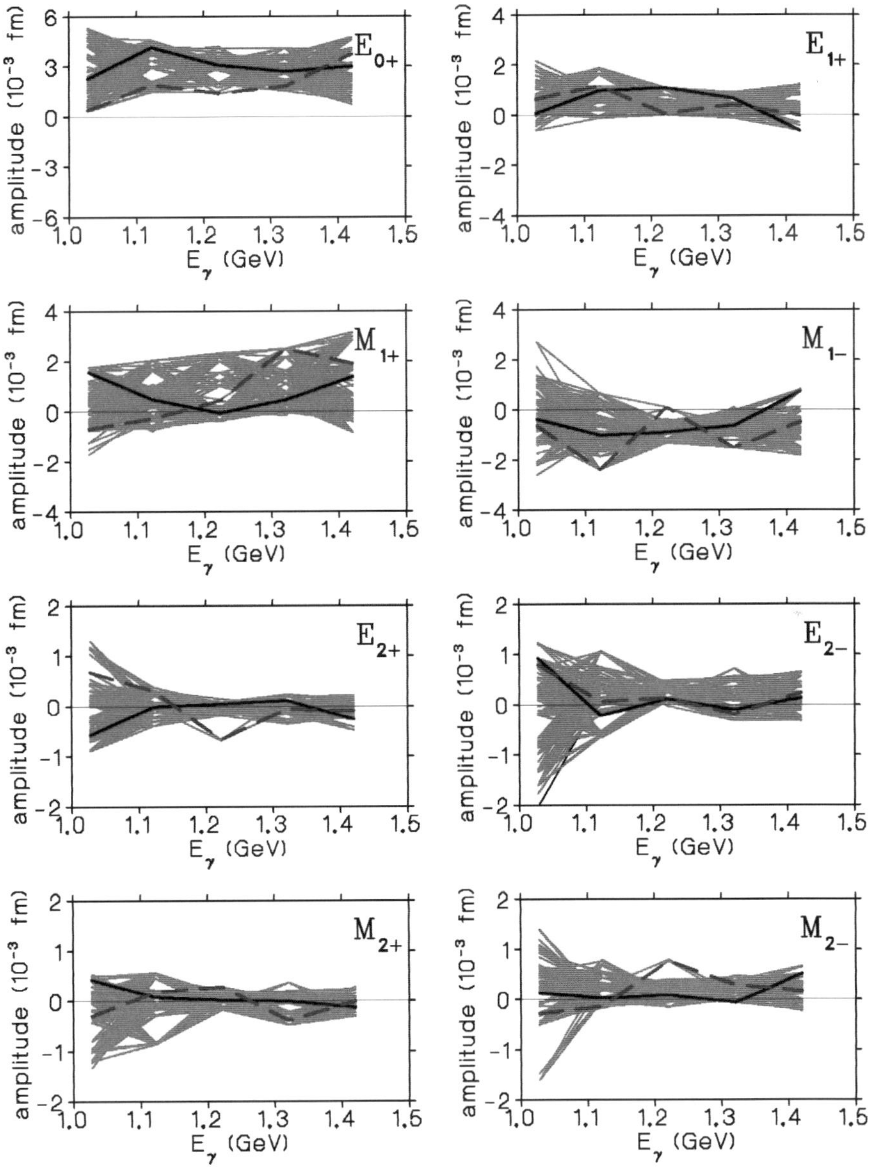

Figure 9. Real parts of the fitted multipoles for L= 0, 1, 2 fitted to published data from [18,19,20,24], with the phase of the E_{0+} multipole fixed to zero [16]. The bands show variations in the χ^2/point of less than 0.2. Solutions with the smallest and largest χ^2/point are shown as the solid (black) and long-dashed (blue) curves, respectively.

be exchanged that are close in mass {K^+(493), K*(892), K_1(1270) and K_1(1400)} and their couplings are not well known. Thus, for the hyperon channel with the greatest potential to supply a complete set of observables, one will always have to fix the phase of one particular multipole.

We have simulated the ultimate data sets we expect to have available in studies of pseudo-data [16]. Asymmetries were generated from a reference multipole solution at the angles and energies of the CLAS measurements, and then dithered about those points by smearing with Gaussians of width given by the expected experimental errors. Multiple local minima are always present, arising essentially from experimental uncertainties. Nonetheless, very significant gains in narrowing of the solution bands come from the addition of new spin observables. Often, even modest errors on an asymmetry are sufficient to derive a substantial gain. It is clear that the criteria for a mathematical solution that have been discussed in the literature [13,23] are unfortunately broken by measurement errors in realistic experiments. Nonetheless, studies with mock-data on a complete suite of 16 observables, accompanied by the uncertainties expected from the recently completed measurements, do result in a dramatic narrowing of the multipole uncertainty bands [16].

7. Summary and outlook

The recent confirmation from LQCD calculations of a large number of as yet unidentified excited nucleon levels has shifted the focus of nucleon structure from speculating why more states might not exist to a concerted effort to find them. Indeed, a failure to locate more than what had been observed in πN scattering would pose a serious problem. New candidates have been proposed, mainly based on recoil polarization asymmetries in hyperon photo-production, although the amplitude solutions remain ambiguous. Given realistically achievable uncertainties, the number of spin asymmetries needed to remove ambiguities in the determination of photo-production amplitudes is many more than the 8 required for an *in principle* mathematical solution. Hyperon channels, particularly KΛ an $K\Sigma^+$ offer the potential for accumulating data on all 16 photo-production spin observables and there now exist large new data sets which are currently under analysis. These are expected to provide a determination of the production amplitudes with minimal uncertainty spread, and these in turn will form the ultimate starting point in a search for new N* poles by analytic continuation. We expect this to create a watershed in our understanding of the nucleon spectrum.

Acknowledgments

The author wishes to thank Drs. T-S. Harry Lee, Sam Hoblit and Craig Roberts for many useful discussions. This work was supported by the US Department of Energy, Office of Nuclear Physics Division, under contract DE-AC05-06OR23177 under which Jefferson Science Associates operate Jefferson Laboratory.

References

[1] Bhagwat M S, Pichowsky M A, Roberts C D and Tandy P C, 2003 *Phys. Rev.* **C68** 015203
[2] Capstick S and Roberts W, 1998 *Phys. Rev.* **D58** 074011
[3] Anselmino M *et al.*, 1993 *Rev. Mod. Phys.* **65**, 1199; and references therein
[4] Maris P, 2004 *Few Body Systems* **35**, 117
[5] Edwards R G, Dudek J J, Richards D G and Wallace S J, 2011 *Phys. Rev.* **D84** 074508
[6] Roper L D, 1964 *Phys. Rev. Lett.* **12** 340
[7] Arndt R A, Briscoe W J, Strakovsky I I and Workman R L, 2006 *Phys, Rev.* **C74** 045205
[8] Aznauryan I G, *et al* (CLAS Collaboration), 2009 *Phys. Rev.* **C80** 055203
[9] Mokeev V I, *et al* (CLAS Collaboration), 2012 *Phys. Rev.* **C86** 035203, and references therein
[10] Li Z, Burkert V and Li Z, 1992 *Phys. Rev.* **D46** 70
[11] Julia-Diaz B, *et al* (EBAC), 2008 *Phys. Rec.* **C77** 045205
[12] Suzuki N *et al* (EBAC), 2010 *Phys. Rev. Lett.* **104** 042302
[13] Chaing W-T and Tabakin F, 1997 *Phys. Rev.* **C55** 2054

[14] The g9-Frost series of experiments, < http://clasweb.jlab.org/shift/g9/>
[15] The g14-HDice experiment, < http://www.jlab.org/exp_prog/proposals/06/PR-06-101.pdf>
[16] Sandorfi A M, Hoblit S, Kamano H and Lee T-S H, 2011 *J. Phys.* G**38**, 053001
[17] Sandorfi A M, Dey B, Sarantsev A, Tiator L and Workman R, arXiv:1108.5411v2
[18] Lleres A *et al* (GRAAL Collaboration), 2007, *Eur. Phys. J.* A**31** 79
[19] McCracken M E *et al* (CLAS Collaboration), 2011 *Phys. Rev.* C**81** 025201
[20] Bradford R K *et al* (CLAS Collaboration), 2006 *Phys. Rev.* C**73** 035202; *ibid*, 2007 C**75** 035205.
[21] Anisovich A V, Beck R, Klempt E, Nikonov V A, Sarantsev A V and Thoma U, 2012, *Eur. Phys. J.* A**48**, 14
[22] Chew G, Goldberger M, Low F and Nambu Y, 1957 *Phys. Rev.* **105** 1345
[23] Barker I, Donnachie A and Storrow J, 1975 *Nucl. Phys.* B**95** 347
[24] Lleres A *et al* (GRAAL Collaboration), 2009, *Eur. Phys. J.* A**39** 149
[25] Workman R L, *priv. comm.*.

Journal of Physics: Conference Series **424** (2013) 012002 doi:10.1088/1742-6596/424/1/012002

Latest KLOE result on the $\sigma(e^+e^- \to \pi^+\pi^-)$ and its impact on the muon anomaly

Giuseppe Mandaglio on behalf of the KLOE/KLOE-2 Collaboration‡

E-mail: gmandaglio@unime.it
Centro di Fisica Nucleare e di Struttura della Materia, I-95125 Catania, Italy
Dipartimento di Fisica, Università di Messina, I-98166 Messina, Italy
INFN, Sezione di Catania, I-95123 Catania, Italy

Abstract. The KLOE experiment was the first to exploit Initial State Radiation (ISR) processes to obtain the $e^+e^- \to \pi^+\pi^-(\gamma)$ cross section below 1 GeV (70% of the leading order contribution to the muon anomaly) publishing two measurements with the photon in the initial state emitted at small angle in 2005 and 2008, and an independent measurement with the photon emitted at large angle in 2011. These measurements were normalized using luminosity from Bhabha. Here, we present the latest KLOE results on $e^+e^- \to \pi^+\pi^-(\gamma)$ cross section, obtained by the ratio $\sigma(e^+e^- \to \pi^+\pi^-)/\sigma(e^+e^- \to \mu^+\mu^-)$ measurement for a total integrated luminosity of 239.2 pb^{-1}. The comparison of the present results with with the previous ones performed by KLOE shows a good agreement, while they are in substantially agreement with the other ones obtained by CMD-2, SND and BaBar experiments. From the cross section we extract the pion form factor and the two-pion contribution in the $\sqrt{s'}$ invariant mass range of 592-975 MeV to the muon anomaly confirming the current discrepancy between standard model calculation and experimental measurement.

1. Introduction

The muon anomalous magnetic moment is one of the most precisely measured quantities in particle physics and the actual discrepancy between standard model (SM) prediction and the experimental measurement arouses great attention and expectations of the international scientific community, because such difference between measurement and

‡ The KLOE-2 Collaboration: D. Babusci, D. Badoni, I. Balwierz-Pytko, G. Bencivenni, C. Bini, C. Bloise, F. Bossi, P. Branchini, A. Budano, L. Caldeira Balkeståhl, G. Capon, F. Ceradini, P. Ciambrone, F. Curciarello, E. Czerwiński, E. Danè, V. De Leo, E. De Lucia, G. De Robertis, A. De Santis, A. Di Domenico, C. Di Donato, R. Di Salvo, D. Domenici, O. Erriquez, G. Fanizzi, A. Fantini, G. Felici, S. Fiore, P. Franzini, P. Gauzzi, G. Giardina, S. Giovannella, F. Gonnella, E. Graziani, F. Happacher, L. Heijkenskjöld, B. Höistad, L. Iafolla, M. Jacewicz, T. Johansson, A. Kupsc, J. Lee-Franzini, B. Leverington, F. Loddo, S. Loffredo, G. Mandaglio, M. Martemianov, M. Martini, M. Mascolo, R. Messi, S. Miscetti, G. Morello, D. Moricciani, P. Moskal, F. Nguyen, A. Passeri, V. Patera, I. Prado Longhi, A. Ranieri, C. F. Redmer, P. Santangelo, I. Sarra, M. Schioppa, B. Sciascia, M. Silarski, C. Taccini, L. Tortora, G. Venanzoni, W. Wiślicki, M. Wolke, J. Zdebik

calculation could be a signal of new physics. In fact, the measurements of the muon magnetic anomaly $a_\mu = (g_\mu - 2)/2$ performed at the Brookhaven Laboratory have reached a remarkable accuracy of 0.54 ppm: $a_\mu = (11\,659\,208.0 \pm 6.3) \times 10^{-10}$ [1], differing from Standard Model of about 3.2-3.6 standard deviations as estimated in Refs. [2–5]. A recent evaluation [6] finds a difference between 4.7 and 4.9 standard deviations. The authors of Ref. [7] have proposed an interpretation in terms of Supersymmetry, which can be probed at the Large Hadron Collider, so far without success. Another proposal suggests the existence of a light vector boson in the Dark Matter sector, coupled with ordinary fermions through photon exchange, which is not excluded by present low energy tests of the Standard Model [8, 9]. A new round of measurements of a_μ will soon begin at Fermilab[10] and JPARC[11], with the aim to considerably reduce the experimental error. To fully exploit the significance of improved measurements it is important to confirm the present estimate of the hadronic corrections and possibly to decrease the corresponding error.

The main source of uncertainty in the Standard Model estimates of a_μ [2, 3, 5] is due to hadronic loop contributions which are not calculable in perturbative QCD. To lowest order, the hadronic contribution $\Delta^{\mathrm{h,lo}}a_\mu$, can be obtained from a dispersion integral [12, 13] over the "undressed" cross section $\sigma^0(e^+e^- \to \mathrm{hadrons}(\gamma))$. The $e^+e^- \to \pi^+\pi^-(\gamma)$ process contributes some 75% of the $\Delta^{\mathrm{h,lo}}a_\mu$ value and accounts for about 40% of the correction uncertainty.

In the following, we present the latest KLOE results on the $\sigma(e^+e^- \to \pi^+\pi^-)$ cross section and the impact on the $\Delta^{\mathrm{h,lo}}a_\mu$ determination.

2. Measurement of $\sigma_{\pi\pi}$ with ISR

The KLOE detector operates at the DAΦNE e^+e^- collider of the INFN Laboratori Nazionali di Frascati. It consists (Fig. 1) of a huge drift chamber providing high momentum resolution($\sigma_p/p \leq 0.4\%$) [14] on reconstructed tracks and Pb-scintillating fibers calorimeter with excellent time ($\sigma_t \sim 54$ ps$/\sqrt{E\,[\mathrm{GeV}]} \oplus 100$ ps) and good energy ($\sigma_E/E \sim 5.7\%/\sqrt{E\,[\mathrm{GeV}]}$) resolution [15].

DAΦNE is a $\phi - $ *factory* running at $\sqrt{s} \simeq M_\phi$, and has delivered ca. 2.5 fb^{-1} of integrated luminosity to the KLOE experiment. So far KLOE has reported two measurements of the $\pi^+\pi^-$ cross section between 0.35 and 0.95 GeV^2 called KLOE05 [16] and KLOE08 [17] in the following. In addition, about 250 pb^{-1} of data have been collected at $\sqrt{s} \simeq 1$ GeV, 20 MeV below the ϕ-meson resonance, from which a new measurement of $\pi^+\pi^-$ cross section was obtained (KLOE10 [18]). Running at energies below the ϕ-meson resonance reduces considerably the background from the copious ϕ-meson decay products, including scalar mesons. The total cross section $\sigma_{\pi\pi} \equiv \sigma_{e^+e^-\to\pi^+\pi^-}$ as a function of s' (the e^+e center-of-mass energy squared after ISR emission) is extracted from the differential cross section $\mathrm{d}\sigma(e^+e^- \to \pi^+\pi^-\gamma)/\mathrm{d}s'$

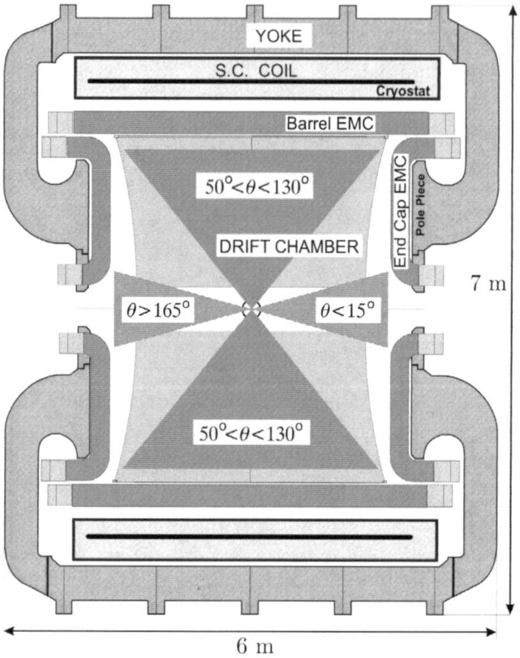

Figure 1. Schematic view of the KLOE detector with selection regions.

using formula [19]:

$$s \cdot \frac{\mathrm{d}\sigma_{\pi\pi\gamma}}{\mathrm{d}s'} = \sigma_{\pi\pi}(s')\ H(s',s)\ , \tag{1}$$

with s the squared e^+e^- center of mass energy, and H the radiator function obtained from theory describing the photon emission in the initial state. An alternative method to extract the $\pi^+\pi^-$ cross section uses the $\pi^+\pi^-\gamma/\mu^+\mu^-\gamma$ ratio [20]:

$$\sigma_{\pi\pi(\gamma)} = \frac{4\pi\alpha^2}{3s'}(1+2m_\mu^2/s')\beta_\mu\ \frac{\mathrm{d}\sigma_{\pi\pi\gamma}/\mathrm{d}s'}{\mathrm{d}\sigma_{\mu\mu\gamma}/\mathrm{d}s'}, \tag{2}$$

with m_μ the muon mass, β_μ the muon velocity in the center-of-mass frame, $\mathrm{d}\sigma_{\pi\pi\gamma}/\mathrm{d}s'$, $\mathrm{d}\sigma_{\mu\mu\gamma}/\mathrm{d}s'$ the $e^+e^- \to \pi^+\pi^-\gamma$, $e^+e^- \to \mu^+\mu^-\gamma$ differential cross sections, respectively. In both Eqs.(1) and (2) the Final State Radiation (FSR) terms are neglected, but are taken into account properly in the analyses.

3. Measurement of the pion form factor from the $\pi\pi\gamma/\mu\mu\gamma$ ratio

The same sample as used for KLOE08 measurement, is analyzed with the small angle photon selection. Equation (2) has been used to extract the pion form factor via a bin-by-bin ratio between the observed pion and muon ISR differential cross sections. This approach has several benefits especially concerning radiative corrections. The integrated luminosity as well as the radiation function H and the vacuum polarisation cancel completely in the $\pi\pi\gamma/\mu\mu\gamma$ ratio. However, the ratio of the acceptance enters in eq. (2) giving corrections of the order of few percents. The additional analysis on

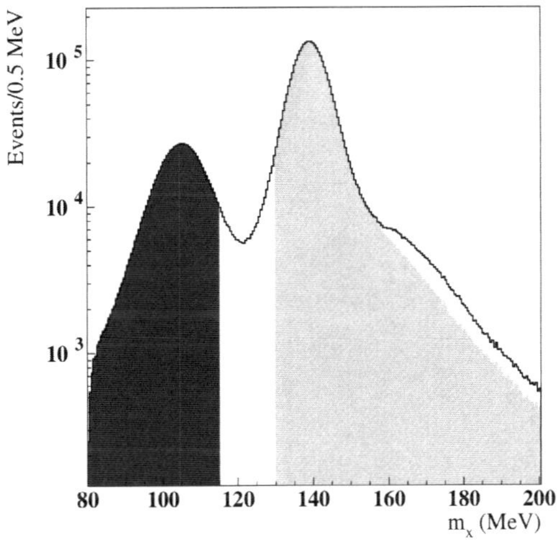

Figure 2. Discrimination of the $\mu\mu\gamma$ (filled left histogram) from the $\pi\pi\gamma$ (filled right histogram) events on the m_x variable.

muon events has to be performed at subpercent level, not a trivial task especially for the pion/muon separation. The analysis for $\pi\pi\gamma$ is essentially the same as for KLOE08 [17], while the one for $\mu\mu\gamma$ is completely new and just for its relevance will be briefly described below.

The signature for $e^+e^-\to\mu^+\mu^-\gamma$ events with the photon emitted at a small angle is just two tracks of opposite curvature, the photon being lost in the beam pipe. Four types of events contribute to the above signature: 1: $e^+e^-\to\mu^+\mu^-\gamma$ 2: $e^+e^-\to\pi^+\pi^-\gamma$ 3: $e^+e^-\to e^+e^-\gamma$, and 4: $e^+e^-\to\pi^+\pi^-\pi^0$. The four reactions can be distinguished kinematically. For the reaction $e^+e^- \to x^+x^-\gamma$, which is once kinematically overdetermined, we can compute the common mass of particles x^+ and x^- by using the following equation: $(\sqrt{s} - \sqrt{|\vec{p}_{x^+}|^2 + m^2_{x^+}} - \sqrt{|\vec{p}_{x^-}|^2 + m^2_{x^-}})^2 - (\vec{p}_{x^+} + \vec{p}_{x^-})^2 = 0$. The four processes give $m_x = m_\mu$, m_π, m_e and $> m_\pi$, respectively. Additional $e\mu$ (and $e\pi$) separation is obtained from a PID estimator L which uses time of flight information and the value and shape of the energy deposits of the two charged particles, of known momentum, in the calorimeter [17]. Fig. 1 shows the fiducial volume we use for muons and unobserved photons which is identical to that used in ref. [17] for $e^+e^-\to\pi^+\pi^-\gamma$. The computed mass for the two observed particle must satisfy $80 < m_x < 115$ MeV (see Fig. 2).

The $\mu\mu\gamma$ cross section measurement has been performed by subtracting the residual background $N_{\rm bkg}$(evaluated by fitting the m_x spectrum of the accepted events with a superposition of Monte Carlo simulation distribution describing signal and concurrent channels, see for more detail [21]) to the observed event count $N_{\rm obs}$ and dividing it for the selection efficiency $\epsilon(M^2_{\mu\mu})$ and integrated luminosity $\mathcal{L}$:

$$\frac{d\,\sigma_{\mu\mu\gamma}}{d\,M^2_{\mu\mu}} = \frac{N_{\rm obs} - N_{\rm bkg}}{\Delta M^2_{\mu\mu}} \frac{1}{\epsilon(M^2_{\mu\mu})\,\mathcal{L}}. \tag{3}$$

The luminosity was evaluated, by using Bhabha events at large angle, to be 239.2

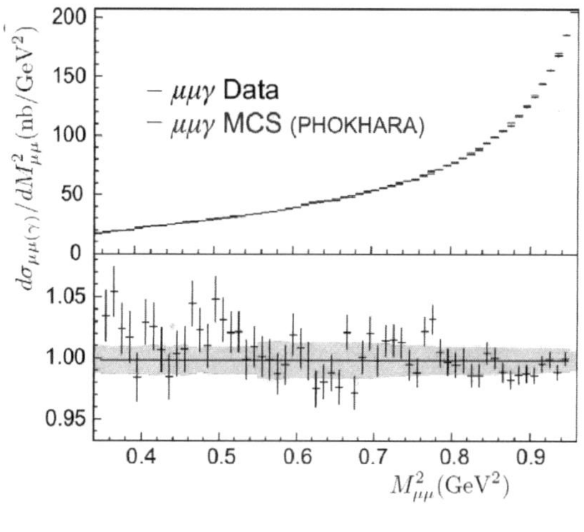

Figure 3. Top: comparison of data (black points) and MCS (blue points) for the $\mu\mu\gamma$ absolute cross section as a function of $M^2_{\mu\mu}$ (GeV^2). Bottom: data to MC ratio. The green band shows the systematic error.

pb^{-1} with a systematic error of 0.3 %[22]. The upper panel of figure 3 shows in the comparison between the measured $\mu\mu\gamma$ cross section and the QED NLO calculation using the PHOKHARA MC code [20]. The lower panel shows the ratio between the two differential cross section. The average ratio (by using statistical error only) of 0.9981 ± 0.0015 shows a good agreement between the experiment and the theoretical evaluations within the estimated uncertainties.

The present analysis (KLOE12) has been checked by using different constraints on the events selection, such as a kinematic fit or tighter cuts on the quality of the charged tracks, all giving to consistent results[21]. Trigger, particle identification and tracking efficiencies have been obtained with data control samples[21].

From the ratio between the $\pi^+\pi^-\gamma$ data published in Ref. [17] and the $\mu^+\mu^-\gamma$ differential cross section described above, we obtain the $e^+e^-\to\pi^+\pi^-$ bare cross section presented in Fig. 4 (inclusive of FSR, VP effects removed):

$$\sigma^0_{\pi\pi(\gamma)}(s') = \sigma(\pi^+\pi^-, s')\,(1+\eta_{\rm FSR})/\,(1-\delta_{\rm VP}) \tag{4}$$

The pion form factor has been then extracted by using the following formula:

$$|F_\pi(s')|^2 = \frac{3}{\pi}\frac{s'}{\alpha^2\beta^3_\pi}\sigma^0_{\pi\pi}(s')(1+\delta_{VP})(1-\eta_\pi(s')) \tag{5}$$

The table summarizing bin-by-bin the results on the bare cross section and the pion form factor are presented in Ref. [21].

The KLOE12 pion form factor compared with the one from KLOE10 and KLOE08, showing good agreement in both cases (see Fig. 5). As one can see in the comparisons reported in Fig. 6, KLOE12 are substantially in agreement with the measurements of CMD-2[23, 24], SND[25] (above the ρ-peak KLOE12 is slightly lower), and BaBar[26, 27] (above 0.6 GeV BaBar results are higher by 2-3%) experiments (see also the comparison with KLOE10 reported in Figs. 4 and 5 of Ref. [18]). The new measurement represents

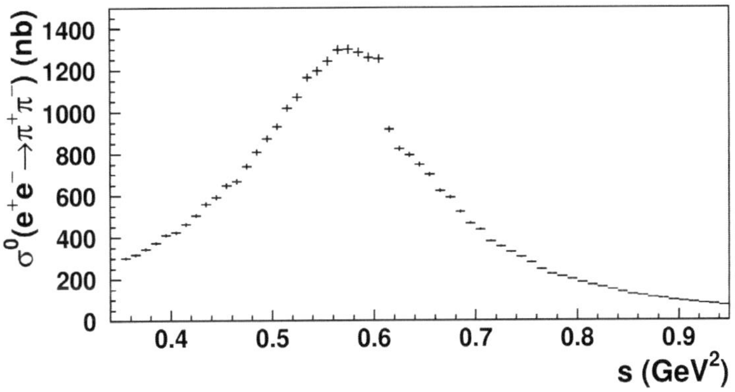

Figure 4. The bare cross section from the $\pi^+\pi^-\gamma/\mu^+\mu^-\gamma$ ratio.

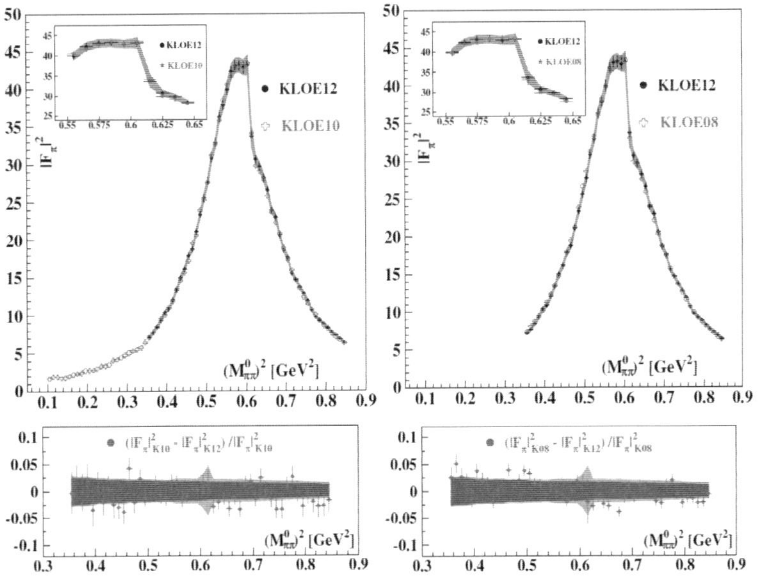

Figure 5. Upper panels: pion form factor obtained with $\pi\pi\gamma$ /$\mu\mu\gamma$ ratio KLOE12 (crosses) compared to the previous KLOE10 (left) and KLOE08 (right) analyses (points). Lower panels: the fractional difference of the new measurement and the previous ones.

an important check of previous measurements, because this last one does not rely on the value of the integrated luminosity and on the radiator knowledge.

The bare cross section $\sigma^0_{\pi\pi(\gamma)}$ is used to determine $a_\mu^{\pi\pi}$ via a dispersion integral:

$$a_\mu^{\pi\pi} = \frac{1}{4\pi^3} \int_{s_{min}}^{s_{max}} \mathrm{d}\, s' \; \sigma^0_{\pi\pi(\gamma)}(s')\, K(s') \; , \tag{6}$$

with s_{min} and s_{max} the lower and upper bounds in the present analysis, and $K(s)$ the kernel function described in [28]. By using Eq. (6), we obtain $\Delta^{\pi\pi} a_\mu = (385.1 \pm 1.1_{\mathrm{stat}} \pm 2.6_{\mathrm{exp}} \pm 0.8_{\mathrm{th}}) \times 10^{-10}$ in the interval $0.35 < M^2_{\pi\pi} < 0.95$ GeV^2. For each $M^2_{\pi\pi}$ bin contributing to the integral, statistical errors are combined in quadrature

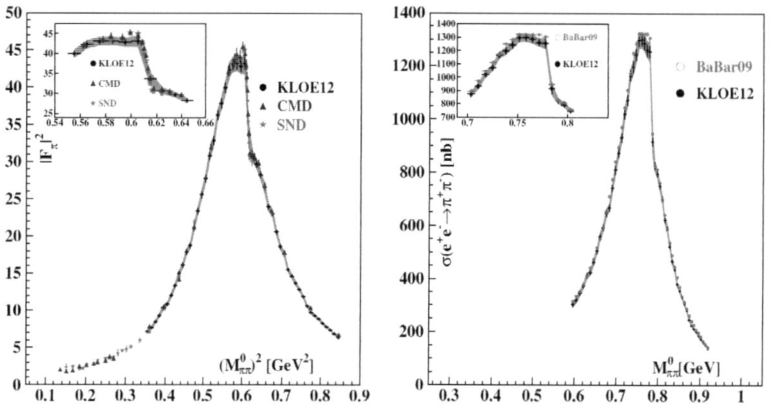

Figure 6. Left panel: pion form factor obtained with $\pi\pi\gamma$ /$\mu\mu\gamma$ ratio KLOE12 (circles) compared to the CMD-2 (triangles) [23, 24] and SND (stars)[25] experiments as a function of $M^2_{\pi\pi}$. Right panel: dipion cross section KLOE12 (full circles) compared to the BaBar (empty circles)[26, 27] experiments as a function of $M_{\pi\pi}$.

and systematic errors are added linearly. A further advantage in the use of the $\pi\pi\gamma$ to

Systematic sources	$\Delta^{\pi\pi}a_\mu$
Background subtraction	0.6%
m_x	0.2%
PID	negligible
Tracking	0.1%
Trigger	0.1%
Unfolding	negligible
Acceptance	negligible
Software Trigger (L3)	0.1%
Experimental systematics	0.7%
Vacuum Polarization	–
FSR resummation	0.2%
Radiator function H	–
Theory systematics	0.2%
Total systematic error	0.7%

Table 1. List of systematic errors on the $\Delta^{\pi\pi}a_\mu$ measurement. Most of them are smaller than the individual uncertainty on $\pi\pi\gamma$ and $\mu\mu\gamma$ due to correlation between the two analyses [21].

$\mu\mu\gamma$ ratio method consists in a reduction of the systematic error in comparison with that published [17] due to correlations between the $\pi\pi\gamma$ and $\mu\mu\gamma$ analyses and a negligible theoretical uncertainty.

The present $a^{\pi\pi}_\mu$ estimation is compared with previous KLOE results (see Table 2).

These results are consistent with the previous estimation thus confirming the current disagreements of about 3σ between the a_μ experimental determination and its Standard Model prediction (see our preliminary estimate in Fig. 7).

Analysis	$\Delta a_\mu^{\pi\pi}(0.35 - 0.85\ \mathrm{GeV}^2) \times 10^{10}$
KLOE12	$377.4 \pm 1.1_{\mathrm{stat}} \pm 2.7_{\mathrm{sys\&theo}}$
KLOE10	$376.6 \pm 0.9_{\mathrm{stat}} \pm 3.3_{\mathrm{sys+theo}}$
	$\Delta a_\mu^{\pi\pi}(0.35 - 0.95\ \mathrm{GeV}^2) \times 10^{10}$
KLOE12	$385.1 \pm 1.1_{\mathrm{stat}} \pm 2.7_{\mathrm{sys\&theo}}$
KLOE08	$387.2 \pm 0.5_{\mathrm{stat}} \pm 3.3_{\mathrm{sys\&theo}}$

Table 2. Comparison between the new result KLOE12 and the published KLOE10 and KLOE08.

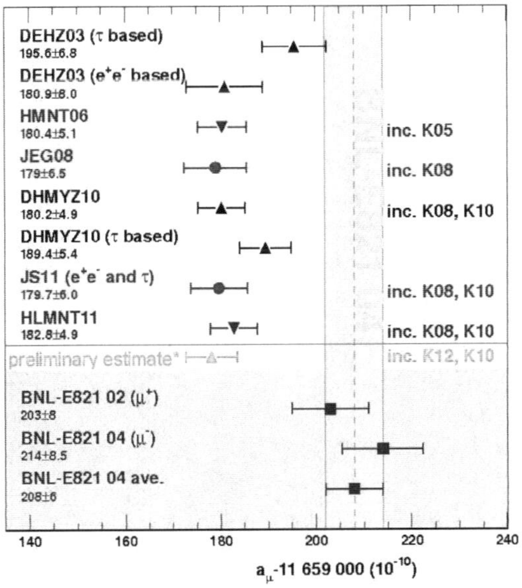

Figure 7. Preliminary estimation based on DHMYZ10 of the muon anomaly including KLOE12 measurement (green full triangle) in comparison with recent theoretical and experimental determinations.

Figure 7 shows our preliminary estimation based on DHMYZ10 (it has not to be considered as a new theoretical estimate) of the muon anomaly a_μ also including the KLOE12 measurement in comparison with recent theoretical and experimental evaluations. The aim of this last figure is to resume the recent status on a_μ estimation an to give an idea of the possible impact of the new KLOE measurements on the present scenario.

4. Conclusions

A evaluation of two-pion cross section, pion form factor, and leading order contribution to muon anomaly by using the $\pi\pi\gamma\ /\mu\mu\gamma$ ratio measurement (KLOE12) has been performed, analyzing the 239.2 pb^{-1} KLOE data. The comparison of this measurement with the previous KLOE08[17] and KLOE10[18] showed a good agreement and represents a powerful check, the $\pi\pi\gamma$ /$\mu\mu\gamma$ result not depending on the integrated luminosity and on the radiator function knowledge. KLOE12 are also in substantially agreement with the measurements performed by the CMD-2[23, 24], SND[25], and BaBar[26, 27] experiments. We also presented the evaluation of $\sigma(e^+e^- \to \mu^+\mu(\gamma))$ normalized to integrated luminosity showing a good agreement with the PHOKHARA MC [20] prediction. The present evaluation of hadronic leading order contribution $a_\mu^{\rm hlo}$ confirms the 3σ discrepancy between SM prediction and experimental measurement [1]. This result stresses on the importance of the expected future improvements on the experimental measurements of the muon anomaly, in program at Fermilab[10] and JPARC[11] laboratories.

References

[1] Bennett G W *et al.* 2006 *Phys. Rev.* D **73** 072003
[2] Jegerlehner F and Nyffeler A 2009 *Phys. Rept* **477** 1
[3] Miller J P, de Rafael E and Roberts B L 2007 *Rept. Prog. Phys.* **70** 795
[4] Davier M, Hoecker A, Malaescu B and Zhang Z 2011 *Eur. Phys. J.* C **71** 1515 [Erratum-ibid. C **72** (2012) 1874]
[5] Hagiwara K, Liao R, Martin A D, Nomura D and Teubner T 2011 *J. Phys. G* **38** 085003
[6] Benayoun M, David P, Buono L D and Jegerlehner F ArXiv:1210.7184 [hep-ph]
[7] Czarnecki A and Marciano W J 2001 *Phys. Rev.* D **64** 013014
[8] Arkani-Hamed N *et al.* 2009 *Phys. Rev.* D **79** 015014
[9] Pospelov M 2009 *Phys. Rev.* D **80** 095002
[10] Carey R M *et al.* Http://lss.fnal.gov/archive/testproposal/0000/fermilab-proposal-0989.shtml
[11] Imazato J 2004 *Nucl. Phys. Proc. Suppl* **129** 81
[12] Bouchiat C and Michel L 1961 *J. Phys. Radium* **22** 121
[13] Lautrup B E and de Rafael E 1971 *Nuovo Cimento* **1** 238
[14] Adinolfi M *et al.* 2002 *Nucl. Inst. Meth.* A **488** 51
[15] Adinolfi M *et al.* 2002 *Nucl. Inst. Meth.* A **482** 364
[16] Aloisio A *et al.* 2005 *Phys. Lett.* B **606** 12 [KLOE Collaboration]
[17] Ambrosino F *et al.* 2009 *Phys. Lett.* B **670** 285 [KLOE Collaboration]
[18] Ambrosino F *et al.* 2011 *Phys. Lett.* B **700** 102 [KLOE Collaboration]
[19] Binner S, Kühn J H and Melnikov K 1999 *Phys. Lett.* B **459** 279
[20] Czyż H, Grzelińska A, Kühn J and Rodrigo G 2005 *Eur. Phys. J.* C **39** 411
[21] Babusci D *et al.* ArXiv:1212.4524 [hep-ex]
[22] Ambrosino F *et al.* 2006 *Eur. Phys. J.* C **47** 589 [KLOE Collaboration]
[23] Akhmetshin R R *et al.* 2007 *Phys. Lett.* B **648** 28
[24] Akhmetshin R R *et al.* 2006 *JETP Lett.* **84** 413
[25] Achasov M N *et al.* 2006 *J. Exp. Theor. Phys.* **103** 380
[26] Aubert B *et al.* 2009 *Phys. Rev. Lett.* **103** 231801 [BaBar Collaboration]
[27] Lees J P *et al.* 2012 *Phys. Rev. D* **86** 032013 [BaBar Collaboration]
[28] Brodsky S and de Rafael E 1968 *Phys. Rev.* **174** 1835

Helicity dependent meson photoproduction on ^{3}He in the Δ-resonance region.

P. Pedroni[1]
For the CBMAMI and A2 Collaborations

[1] INFN, Sezione di Pavia, Via A. Bassi 6, 27100 Pavia, Italy

E-mail: pedroni@pv.infn.it

Abstract. The first measurement of the helicity dependence of the total inclusive photo-absorption cross section and of partial photon-induced reaction channels on ^{3}He was carried out at MAMI (Mainz) in a photon energy range between 200 and 450 MeV. The experiment used the large acceptance Crystal Ball spectrometer, complemented by a charged particle and a vertex detector, a circularly polarized tagged photon beam and a longitudinally polarized high-pressure ^{3}He gas target. The results obtained give information on the GDH sum rule on the neutron and allow an investigation of the modifications of nucleon properties inside ^{3}He.

1. Introduction

Since the beginning of the 1960s, a central issue of nuclear and particle physics has been the determination of the internal structure of the nucleon, in particular of the spin structure, which is not so well understood as other nucleon properties. For this study, of particular interest are sum rules, which connect information from all energies to fundamental parameters of the present interaction models. The Gerasimov-Drell-Hearn sum rule [1, 2] is a good example of these rules. It relates the nucleon anomalous magnetic moment (AMM) κ , the spin S and the mass M of a nucleon to the integral over the weighted helicity difference of the total absorption cross section for circularly polarized photons on a longitudinally polarized nucleon target. It can be written as:

$$I_{GDH} = \int_{\nu_{th}}^{\infty} \frac{\sigma_p - \sigma_a}{\nu} d\nu = 4\pi^2\kappa^2\frac{e^2}{M^2}S\ , \qquad (1)$$

where ν is the photon energy, α is the fine-structure constant, σ_P and σ_A denote the total absorption cross section for parallel and antiparallel orientation of photon and particle spins, respectively. The lower limit of the integral, ν_0 corresponds to pion production and photodisintegration threshold for a nucleonic and nuclear target, respectively. Table 1 reports the magnetic moment μ and the AMM κ values for protons, neutrons, deuterons and ^{3}He nuclei. The GDH sum rule is derived from very general fundamental physical principles, in particular from the forward Compton scattering, the optical theorem and the low energy theorems. In the past, there have been several attempts to find causes for a failure of the GDH sum rule. The only "weak" hypothesis is the assumption that the Compton scattering becomes spin-independent as the photon energy tends to infinity. Possible explanations for this violation could be the exchange of a_1-like meson between the photon and the nucleon or a non-pointlike quark structure.

Table 1. Magnetic moment μ and AMM κ (units: μ_N, nuclear magneton) for protons, neutrons, deuterons and ^{3}He nuclei.

	p	n	d	^{3}He
μ	2.79	-1.91	0.86	-2.13
κ	1.79	-1.92	0.65	498

The test of this relation then provides a fundamental check of our knowledge of the γ-nucleon interaction, as well as of the physics of strongly interacting systems; it is also a check of the existing photo-reaction models. In addition, through the helicity dependent partial channels, it will be possible to access new observables and to determine the baryon resonance properties.

An estimate of the GDH sum rule value for the nucleon can be evaluated from the existing multipole analyses of the available single pion photoproduction data (mostly from unpolarized experiments) [3, 4] and phenomenological models of multipion and heavy meson photoproduction reactions [5, 6, 7] up to $E_\gamma \simeq 2$ GeV. Above this energy, the contribution can be calculated from Regge-type approaches [8]. Table 2 reports the current theoretical estimates of the GDH sum rule values both for the proton and the neutron. Predictions for the $N\pi$ channels are both from SAID [3] and MAID [4] (within brackets) multipole analyses.

A discrepancy can be clearly seen between these theoretical predictions and the GDH sum rule value for the proton, while the neutron GDH value is roughly reproduced. In addition, it is worth noting that according to the models, I_{GDH} for the proton and the neutron are roughly the same, while this is not the case for the GDH sum rule predictions. In order to find out the reasons for this discrepancy, a precise measurement of the GDH integral for both the proton and the neutron is needed, as well as a systematic study of the partial channels, in particular of the $N\pi(\pi)$ ones, which give the dominant contribution to the GDH integral.

The first experimental check of the GDH sum rule for the proton was performed by the GDH collaboration jointly at the Mainz and Bonn tagged photon facilities, where the I_{GDH} was experimentally evaluated in the photon energy range between 200 MeV and 2.9 GeV [9, 10, 11, 12]. The combination of this result with the theoretical predictions for the unmeasured energy ranges gives an estimated value of the GDH integral of 211 $\pm$ 5(stat) $\pm$ 12(sys)μb. The obtained result supports the validity of the GDH sum rule for the proton, at odds with the theoretical estimates given in Table 2. This discrepancy is mainly due to the oscillating photon energy dependence of the GDH integrand, due to the alternating sign of the multipole contributions. In order to have a reliable prediction of the GDH integral, a very high accuracy, that has not been reached yet, is needed for the theoretical models.

2. The GDH sum rule for the neutron

In the neutron case, the experimental verification of the GDH sum rule is complicated by the lack of free neutron targets. In order to compensate for this lack, deuteron or high pressure ^{3}He targets can be used. In both cases, nuclear structure effects and final state interactions prevent the direct access to the the free nucleon cross section of and the evaluation of the neutron contribution will be model dependent. The first experimental measurement carried out using longitudinally polarized deuterons has been performed in the energy region between 0.2 and 1.8 GeV by the GDH collaboration [13, 14, 15]. In [15] a very rough estimate was derived for the GDH integral value for the neutron from the combination of these data with the ones from the proton. However, the lack of reliable nuclear models describing in a satisfactory manner the helicity dependent γ -d interactions and the presence of a large proton background contribution,

Table 2. Theoretical estimates of the GDH sum rule values for the proton and the neutron.

	I_{GDH} proton	I_{GDH} neutron
$\gamma N \rightarrow N\pi$	172 [164]	147 [131]
$\gamma N \rightarrow N\pi\pi$	94	82
$\gamma N \rightarrow N\rho$	-8	-6
$\gamma N \rightarrow K\Lambda(\Sigma)$	-4	2
$\gamma N \rightarrow N\rho(\omega)$	0	2
Regge contribution ($E_\gamma > 2$ GeV)	-14	20
TOTAL	$\sim$ 239 [231]	$\sim$ 247 [231]
GDH sum rule	204	233

prevent at the moment a reliable extraction of the GDH neutron value.

A complementary and more direct access to a free polarized neutron is given by a longitudinally polarized ^{3}He target. While the proton and the neutron inside the deuteron are essentially in s-states of relative motion with aligned spins, ^{3}He is (with $\sim$ 90% probability) a system consisting of two protons with spins paired off and an active unpaired neutron, in relative s-states. As a result, the spin contribution of the two protons cancels off and the magnetic moment of ^{3}He can be approximated with the magnetic moment of the neutron. Therefore, it is expected that the measured GDH integral for ^{3}He above the pion production threshold will already be a good approximation of the GDH integral value for the neutron. In a PWA approach, and above the pion production threshold, the following formula can then be used:

$$I_{GDH}^{^3He} = p_n \cdot I_{GDH}^{neutron} + p_p \cdot I_{GDH}^{proton} \tag{2}$$

where $p_p = -0.052$ and $p_n = 0.87$ are the are the effective degrees of neutron and proton polarization inside ^{3}He as evaluated by [16]. Since the proton contribution to the measured helicity dependent yields is much smaller than in the deuteron case(for which $p_p = p_n = 0.92$), it can be clearly seen that the most accurate evaluation of I_{GDH} will come from ^{3}He.

In any case, the comparison between the two different free neutron values extracted from both targets using different nuclear models will play a crucial role in constraining the theoretical analyses and will give a fundamental cross check of the reliability of the free neutron extraction procedures.

3. Experimental Set-up

The experiment was carried out at the tagged photon facility of the MAMI accelerator in Mainz. Circularly polarized photons were obtained by bremsstrahlung of longitudinally polarized electrons having an energy of 525 MeV and an average polarization of about 80%. The relative electron polarization was continuously monitored using a Moeller polarimeter and its absolute value was periodically measured using a Mott polarimeter. This parameter was then determined with an absolute accuracy of 3%. The bremsstrahlung photons were tagged using the Glasgow-Mainz magnetic spectrometer with an energy resolution of about 1 MeV [17, 18, 19]. The relative tagging efficiency was monitored throughout the experiment using an ionization chamber which measures the overall photon flux and absolute measurements were regularly made by a total-absorption lead glass counter, which was moved into the beam line at reduced photon intensity.

In this way, the intensity of the tagged photon flux was known with an accuracy of 5% [20].

A high-pressure (~4 bar) polarized ^{3}He gas target has for the first time been used with a photon beam line. The polarized gas was contained in a cylindrical cell with a total length of 20 cm and an outer diameter of 6 cm. It is made from quartz glass with two 50 μm thick titanium foils as entry and exit windows for the photon beam (see figure 1). These materials were chosen since they provide the necessary gas tightness and give an acceptably long relaxation time of the gas polarization. Under these conditions, the ^{3} He gas target density is relatively low ($\approx 2.5 \cdot 10^{21}$ nuclei/cm^2) compared to that of a solid or liquid target. Despite this, the ^{3}He gas target is pure, so it has a greater fraction of polarized neutron with respect to the deuterated butanol case.

After the polarization process of the gas, performed outside the experimental area using the the Metastability Exchange Optical Pumping (MEOP) method [16], the target cell was inserted into the detector system. There a solenoid provided a very homogeneous guiding magnetic field to maintain the polarization alignment (see figure 2). A relative measurement of the polarization was done every hour using NMR techniques. The principles of operation for this target, as well as the complete target setup used in the experiment are detailed in [22]. The ^{3}He nuclei were polarized typically up to about 70% with relaxation times of about 20 hours. The target density and polarization degree were known with an accuracy of 2% and 5% respectively [22].

The reaction products were detected by a detector system, consisting of the Crystal Ball (CB) NaI spectrometer, complemented by the Multi-Wire Proportional Chambers (MWPCs), used to identify and track the charged particles, and the cylindrical Particle Identification Detector (PID), used to distinguish the charged from the neutral particles detected by the CB (see figure 2). The combined information provided by these three detectors provides accurate energy, angle and particle identification in the azimuthal (ϕ) and polar (θ) angular regions from 0° to 360° and from 21° to 159°, respectively.

Finally, in order to suppress as much as possible the background originating from reactions on the atomic electrons, a threshold Cherenkov detector was installed. The detector was located downstream of the CB detector to cover the polar angular range from 0° to 18°, where nearly all atomic events occur.

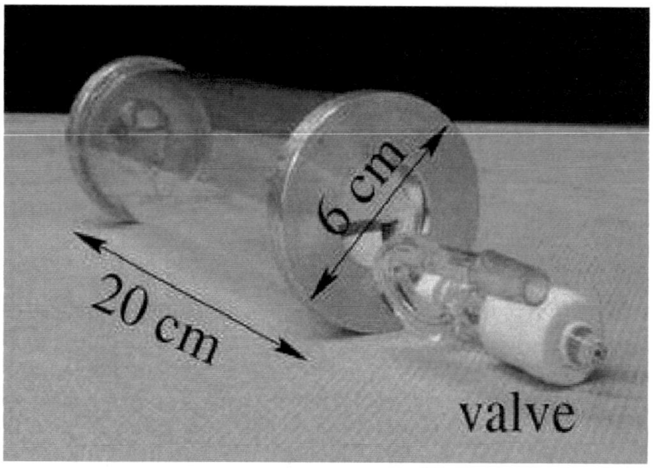

Figure 1. The polarized ^{3}He target cell.

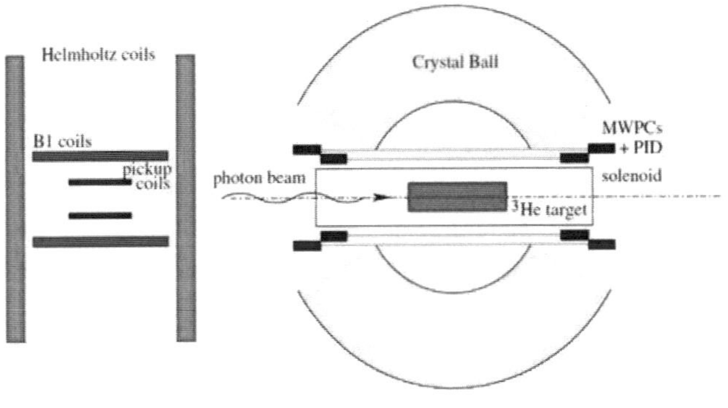

Figure 2. Schematic view of the experimental set-up.

4. Unpolarized data analysis

4.1. Total inclusive cross section (σ_{tot})

To avoid large systematic uncertainties arising from the detection of all single partial reaction channels, an inclusive method of data analysis has been developed to directly determine the total inclusive photoabsorption cross section (see, for instance, [23]). In this method the identification of individual processes is not required; what is necessary is to observe at least one reaction product of all possible hadronic final states with almost complete acceptance, as far as solid angle and efficiency are concerned. The corrections needed to evaluate the detector efficiencies and the loss of events emitted in the angular/momentum regions not covered by the detector have to be kept as low as possible to minimize model dependent contributions. Due to the roughly isotropic distributions of the photo-emitted pions and of the protons coming from photodisintegration in the considered photon energy range, the CB detector, with a very large covered solid angle and an instrinsic detection efficiency greater than about 99% for both charged hadrons and photons coming from neutral meson decays, meets these requirements.

The event selection procedure for the inclusive method was quite simple: at least one cluster (i.e. a group of adjacent hit crystals) was required to be present inside the CB. To further reject e.m. background, only clusters which had a total energy of 40 MeV or higher were used in the analysis. Monte Carlo simulations show that, under these conditions, a large fraction of σ_{tot} (from $\sim$ 90% at E_γ= 200 MeV to $\sim$ 96% at E_γ= 500 MeV) can be directly accessed since, for the dominant quasi-free processes on single nucleons, the minimum pion momenta for the $\pi^{\pm}X$ channels are above the CB detection threshold.

A model dependent extrapolation was evaluated to obtain the remaining part of the total photoabsorption cross section which produces events where all charged hadrons and/or photons from π^0 decay are emitted outside the detector acceptance. The corrections for the $\gamma\ ^3He \rightarrow \pi X$ reactions were evaluated assuming that only quasi-free processes on the single nucleons are present and using the angular distributions for the $\gamma N \rightarrow \pi N$ reactions were evaluated processes predicted by the MAID [4] multipole analysis. The missing contribution from the the $\gamma\ ^3He \rightarrow ppn$ channel has been evaluated taking into account that the dominant reaction mechanism is the absorption on a correlated (n,p) pair. The systematic uncertainty associated to the simplified models used to evaluate the extrapolation corrections is estimated to be 10% of the calculated correction. The combination of all different sources, gives an overall systematic error of about 6% of σ_{tot}.

In figure 3 the values of the unpolarized total inclusive cross section obtained from the present experiment after the subtraction of the empty target spurious contributions [24] are compared to

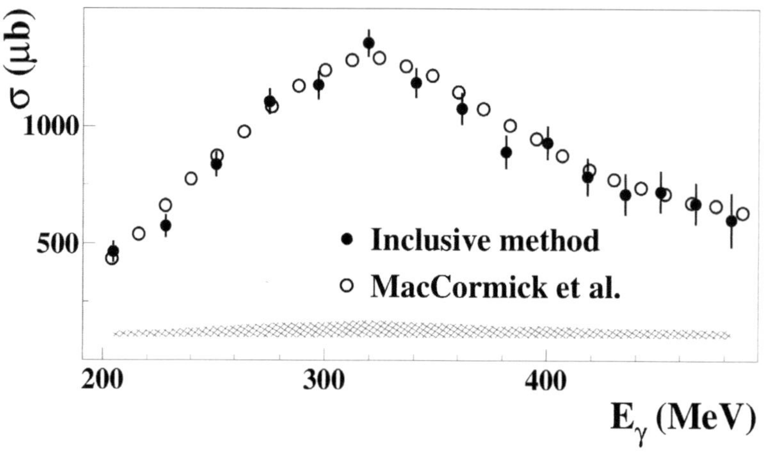

Figure 3. The unpolarized total inclusive photoabsorption cross section on ^{3}He (full circles) is compared to the unpolarized results obtained by the DAPHNE collaboration (open circles) [23].

previous results [23]. The good agreement that can be clearly seen with respect to the published data gives confidence in the total inclusive procedure.

4.2. Partial reaction channels

In order to provide additional experimental information, the particle identification capabilities of the experimental apparatus were used to evaluate the total cross section for the semi-exclusive processes (i) $\gamma\ ^3\mathrm{He} \to \pi^0 X$ ($\sigma_{\pi^0 X}$), (ii) $\gamma\ ^3\mathrm{He} \to \pi^\pm X$ ($\sigma_{\pi^\pm X}$), and for the photodisintegration reaction (iii) $\gamma\ ^3\mathrm{He} \to ppn$ (σ_{ppn}). All these cross sections have not been measured yet.

The yield from (i) was evaluated by selecting events having two or three neutral clusters in the CB and the π^0 mesons were identified by a standard invariant mass analysis. The main parasitic reactions that can contaminate the data are the $\gamma\ ^3\mathrm{He} \to \pi^0\pi^0 X$ channels. For this reason, the data analysis was limited to $E_\gamma < 450$ MeV, a region where this contribution is small. The residual contamination was evaluated assuming the dominance of the quasi-free $\gamma N \to \pi^0\pi^0 N$ reactions and using the measured unpolarized cross section values for these processes previously measured by the GDH [25, 26, 27] and TAPS [28] collaborations.

For this and for the other partial channels, the extrapolation correction have been evaluated as previously explained for the total inclusive method. The systematic uncertainty associated to the event selection procedure and to the efficiency corrections was evaluated to be 3% of the measured yield [20]. The addition in quadrature of all the different sources of systematic uncertainties gives an overall systematic uncertainty of about 7% of the measured reaction yield. The obtained results are shown in figure 4a and compared to the predictions of the Fix-Arenhovel (FA) model (solid line) and of a very simple (VS) model based on the MAID multipole analysis (dashed line) [24]. The FA model is a straightforward extension of the work previously done on the deuteron [30]. The elementary production operator $\gamma N \to \pi N$ is taken from the MAID multipole analysis and is afterwards embedded into the wave function to take into account the nuclear effects. Empirical attenuation factors were then applied to take into account the absorption of the photoemitted particles inside the nuclear medium. On the contrary, in our VS model it naively assumed that all nucleons inside ^{3}He behave as if they were free. As it can be clearly seen from the difference between these two models, the predicted role of the nuclear effects result in damping and broadening the peak corresponding to the $\Delta(1232)$ resonance excitation.

Journal of Physics: Conference Series **424** (2013) 012003 doi:10.1088/1742-6596/424/1/012003

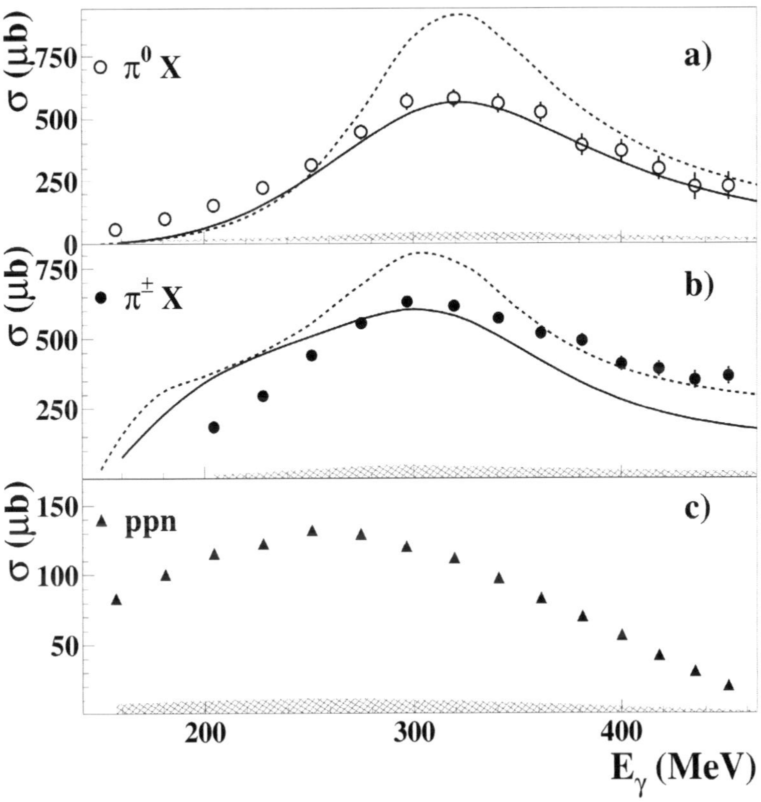

Figure 4. The unpolarized total cross section for a) γ ^{3}He $\to \pi^0 X$ b) γ ^{3}He $\to \pi^\pm X$ and c) γ ^{3}He $\to ppn$. The error bars are statistical and the hatched bands show the systematic uncertainties. In a) and b) the experimental data are compared to the FA model (solid line) and to our VS model (dashed line).

Proton and charged pions hitting the CB were identified by a standard $dE/dX - E$ analysis, using the energy information from CB, PID and the direction information from MWPCs [20]. In both cases, software cuts on the interaction vertex coordinates of the selected events allowed a suppression of a large fraction of the events originating from the target walls and windows [22]. The trajectory reconstruction efficiency was directly determined from the experimental data since a part of the $\pi^\pm$ events can be discriminated using the approximate angular information given by CB alone instead of the one given by MWPCs. This efficiency was then determined by measuring the fraction of such events that have a reconstructed trajectory using MWPCs information and was found to be about 85% (95%) for $\pi^\pm$ (protons) with a smooth dependence on the incident photon energy. An absolute systematic uncertainty of 3% has been estimated on this parameter.

Events from reaction (ii) were obtained by requiring one charged pion being identified in CB. In a similar way as before, the data analysis was limited to $E_\gamma < 450$ MeV and the small contamination due to the γ ^{3}He $\to \pi^+\pi^- X$ channels was evaluated from the $\gamma p \to \pi^+\pi^- p$ cross section measured by the GDH collaboration [29]. The systematic uncertainty associated to the event selection procedure and to efficiency corrections was evaluated to be 3% of the measured

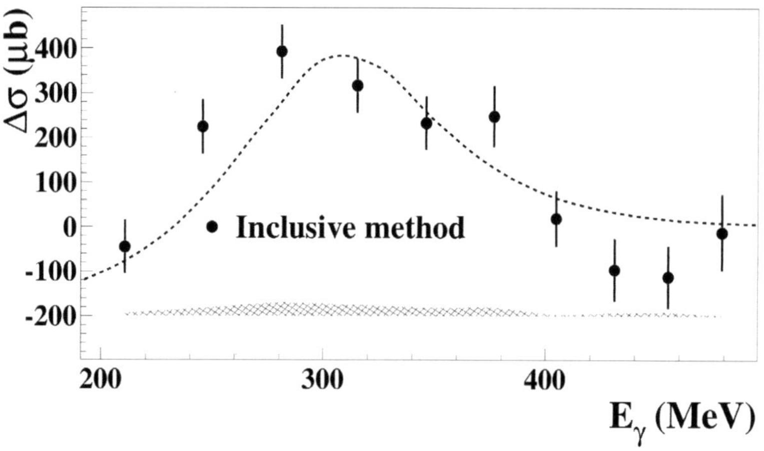

Figure 5. The polarized total inclusive cross section on ^{3}He ($\Delta\sigma_{tot}$) (full circles) compared to the predictions of our VS model (dashed line). The error bars are statistical and the hatched band shows the systematic uncertainties.

yield. The addition in quadrature of all the different sources of systematic uncertainties gives an overall systematic uncertainty of about 7% of the measured reaction yield.

The obtained results are shown in figure 4b and compared to the predictions of the FA model (solid line) and of our VS model [24]. In this case the FA model does not well reproduce our data which for to $E_\gamma > 350$ MeV are very close to our simple model.

The yield from reaction (iii) was evaluated by selecting events having one or two protons identified in CB. Since the major competing background is due to the $\gamma\ ^3\text{He} \to ppn\pi^0$ reaction, the presence of less than two neutral clusters detected inside CB in coincidence with the proton(s) was also required. The residual background was evaluated from the missing mass spectrum $\gamma\ ^3\text{He} \to p(p)X$ and subtracted accordingly. The addition in quadrature of all the different sources of systematic uncertainties gives an overall systematic uncertainty of about 8% of the evaluated reaction cross section. The resulting total unpolarized cross section is shown in figure 4c [24]. In this case no specific model is available for comparison.

5. Polarized results and comments

In the analysis of the helicity dependent data, all previously mentioned analysis methods were used to evaluate the difference $\Delta\sigma = (\sigma_p - \sigma_a)$. In this case the contributions from all non-polarized materials present in the target cell vanish. In a similar way as before, the extrapolation correction for $\gamma\ ^3\text{He} \to \pi X$ was evaluated using the MAID predictions while the $\Delta(pn)$ contribution was calculated using the predictions of the SA model.

The analysis procedure described above results in the helicity dependent total inclusive cross section $\Delta\sigma_{tot}$ as depicted in figure 5 [24] in comparison with the predictions of our VS model, where it was again assumed that the nucleons inside ^{3}He behave as free ones. Only the effects on the nucleon spin alignments due to the ^{3}He s' and d-state probabilities have been taken into account according to equation 2. The agreement between our data and the VS model is reasonable, taking into account the non-negligible statistical experimental errors. This is a hint that nuclear effects are less important than in the unpolarized case.

The total helicity dependent cross sections $\Delta\sigma$ for the (a) π^0X, (b) $\pi^\pm X$ and (c) ppn channels are shown in figure 6 together with the corresponding predictions of the FA (solid line) and VS

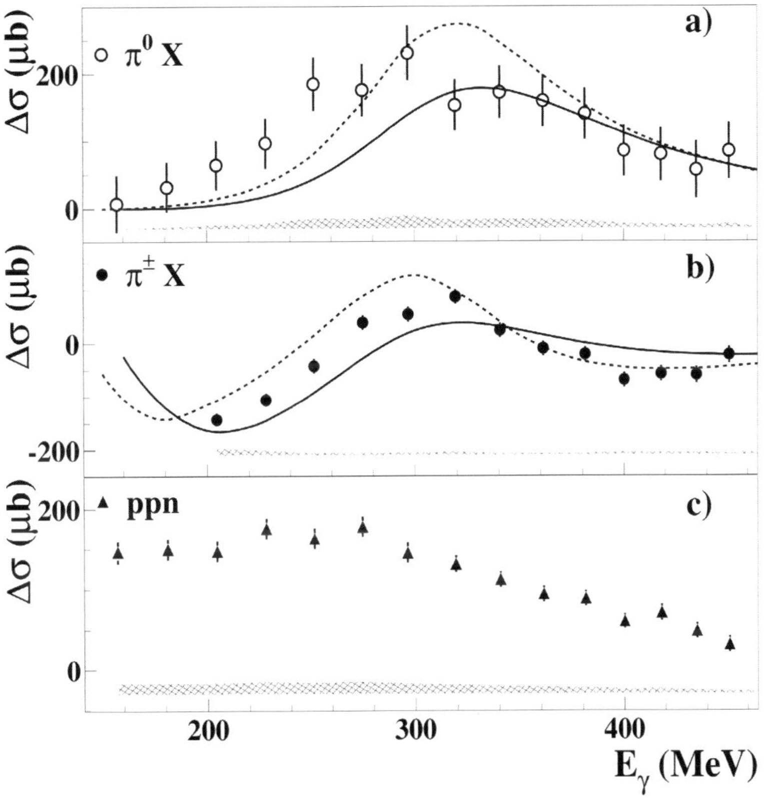

Figure 6. The polarized total cross section for the a) $\gamma\ ^3\mathrm{He} \rightarrow \pi^0 X$ b) $\gamma\ ^3\mathrm{He} \rightarrow \pi^{\pm} X$ and c) $\gamma\ ^3\mathrm{He} \rightarrow ppn$. The error bars are statistical and the hatched bands show the systematic uncertainties. In a) and b) the experimental data are compared to the FA model (solid line) and to our VS (see eq. 2) model (dashed line).

(dashed line) model [24]. As in the unpolarized case the FA model fairly well reproduces the $\gamma\ ^3\mathrm{He} \rightarrow \pi^0 X$ data only at higher photon energies and it does not reproduces the shape of the $\gamma\ ^3\mathrm{He} \rightarrow \pi^{\pm} X$ data. Our VS model well reproduces the data at higher photon energies for both reactions. This is a further confirmation that the effects more directly related to the composite nuclear target structure do not have a strong helicity dependence and their net effect is then reduced in the $\Delta\sigma$ case.

6. Conclusions

The helicity dependence of the total inclusive photoabsorption cross section on ^{3}He and both the unpolarized and the helicity dependent cross sections of the partial reaction channels $\gamma\ ^3\mathrm{He} \rightarrow \pi^0 X$, $\gamma\ ^3\mathrm{He} \rightarrow \pi^{\pm} X$ and $\gamma\ ^3\mathrm{He} \rightarrow ppn$ have been measured for the first time at MAMI (Mainz) in the energy region $200 < E_\gamma < 450$ MeV. All these new data provide a very powerful tool to further improve the models for the photoreactions on ^{3}He in the Δ resonance region. Available state-of-the-art calculations are not able to describe in a s atisfactory manner both the unpolarized and the helicity dependent cross section for the πX channels while no model is at present available for the ppn channel. This fact strongly motivates further theoretical and experimental research in the field. In order to pin down the origin of these discrepancies the analysis of both the unpolarized and polarized differential cross sections for the $\gamma\ ^3\text{He} \to \pi^0 X$ and $\gamma\ ^3\text{He} \to \pi^\pm X$ is underway.

References

[1] Gerasimov S B 1966 *Sov. J. Nucl. Phys.* **2** 430
[2] Drell S D and Hearn A C 1966 *Phys. Rev.* **96** 1428
[3] Arndt R A et al 2002 *Phys. Rev.* C **66** 055213
[4] Drechsel D et al 2001 *Phys. Rev.* D **63** 114010
[5] Fix A and Arenhövel H 2005 *Eur. Phys. J.* A **25** 115
[6] Sumowigado S and Mart T 1999 *Phys. Rev.* C **60** 028201
[7] Zhao Q, Al-Khalili J S and Bennhold C 2002 *Phys. Rev.* C **65** 032201
[8] Bianchi N and Thomas E 1999 *Phys. Lett.* B **450** 439
[9] Ahrens J et al 2000 *Phys. Rev. Lett.* **84** 5950
[10] Ahrens J et al 2001 *Phys. Rev. Lett.* **87** 022003
[11] Dutz H et al 2003 *Phys. Rev. Lett.* **91** 192001
[12] Dutz H et al 2004 *Phys. Rev. Lett.* **93** 032003
[13] Dutz H et al 2005 *Phys. Rev. Lett.* **94** 162001
[14] Ahrens J et al 2006 *Phys. Rev. Lett.* **97** 202303.
[15] Ahrens J et al 2009 *Phys. Lett.* B **672** 328
[16] et al 1990 *Phys. Rev.* C **42** 2310
[17] Anthony I et al 1991 *Nucl. Instrum. Methods Phys. Res., Sect.* A**301**, 230
[18] Hall S J et al 1996 *Nucl. Instrum. Methods Phys. Res., Sect.* A**368**, 698
[19] McGeorge J C et al 2008 *Eur. Phys. J.* A **37** 129
[20] Schumann S et al 2010 *Eur. Phys. J.* A **43** 269
[21] Colegrove F et al 1963 *Phys. Rev.* **132** 2561
[22] Krimmer J et al 2011 *Nucl. Instrum. Methods Phys. Res., Sect.* A**648**, 35
[23] Mc-Cormick et al 1996 *Phys. Rev.* C **53** 41
[24] Aguar-Bartolomé et al, submitted to *Phys. Lett.*B
[25] Ahrens J et al 2003 *Phys. Lett.*B **551** 49.
[26] Ahrens J et al 2005 *Phys. Lett.*B **624** 173.
[27] Ahrens J et al 2011 *Eur. Phys. J.* A **47** 36.
[28] Kotulla M et al 2005 *Phys. Lett.*B **578** 63.
[29] Ahrens J et al 2007 *Eur. Phys. J.* A **34** 11.
[30] Fix A and Arenhövel H 2005 *Phys. Rev.* C **72** 064004

Particle production in ultra-relativistic proton-proton and heavy ion collisions at the Large Hadron Collider

Francesco Riggi
Department of Physics and Astronomy and INFN, Sezione di Catania, Via S.Sofia 64, I95123 Catania, Italy
E-mail: Francesco.Riggi@ct.infn.it

Abstract.
Recent results concerning the main features of particle production in proton-proton and heavy ion collisions at the energy regime of the Large Hadron Collider, as recently made available from the ALICE, ATLAS, CMS and LHCb Collaborations, are discussed.

1. Introduction

The production of hadronic particles in nuclear, particle and cosmic-ray physics is of special interest to understand the underlying mechanisms leading to such reaction products and to test the predictions from non-perturbative QCD processes. The yield of identified hadrons, their multiplicity distributions, as well as the rapidity and transverse momentum spectra are the basic observables in proton-proton and heavy ion collisions at any energy regime, from a few GeV per nucleon to the new ultra-relativistic LHC regime, spanning c.m. energies of a few TeV.

The measurement of such global observables when entering a new energy regime has produced over the past years, since the beginning of the RHIC era (c.m. energies up to 200 GeV), new and interesting insights about the involved production mechanisms, allowing in turn to strongly improve the theoretical description of such processes and tune the corresponding event generators. The study of particle production is also an important tool to provide a description of the expected background for the measurements of hard and rare interactions.

In pp collisions at ultra-relativistic energies the bulk of the particles produced at mid-rapidity have transverse momenta below 1 GeV/c. First principles calculations based on perturbative QCD are not able to provide detailed predictions of particle production. The experimental measurements of the basic properties of particle production is thus essential to provide reliable input for hadronic model calculations. Moreover, a detailed investigation of identified hadrons as a function of the transverse momentum is required.

2. The new energy regime at the LHC

The Large Hadron Collider (LHC) at CERN started its physics program on November 2009, providing the first proton beams at $\sqrt{s}$=0.9 TeV. Since then, an increasing amount of physics data has been collected by all running experiments, by the study of pp, PbPb and pPb collision events.

The study of heavy ion collisions at high bombarding energies is now a mature field. After the pioneering experiments in Berkeley and Dubna in the 70's with relativistic heavy ions, the first experiments at ultra-relativistic energies date to 1986 with light ions at the Brookhaven AGS and the CERN SPS. Heavier projectiles became available in the AGS since the end of 1992 and at the SPS since the end of 1994. Only a few years later, in 2000, with the opening of the Relativistic Heavy Ion Collider (RHIC), the first Au-Au collisions at a centre-of-mass energy of 130 GeV were studied. Since then, a variety of new physics signals and results have been obtained at RHIC. As a result of these experiments, a first understanding of the new high energy density state created in such energetic collisions as an extremely strongly interacting and almost perfect fluid has emerged. According to experimental evidence obtained so far at RHIC, one of the properties of such medium is of being able to absorb much of the energy of fast partons (quarks and gluons) traveling through it, a process referred to as jet quenching. Moreover, it exhibits little internal friction, i.e. very small viscosity.

The first proton-proton collisions at the new Large Hadron Collider were observed in November 2009, and only one year later, in November 2010, the first Lead ion beams collided at a c.m. energy of 2.76 TeV, providing a large data set on nucleus-nucleus collisions in a new energetic regime. In about 25 years, the available energy in the center-of-mass system has thus increased by four orders of magnitudes, giving access to new physics and unexplored regions of the nuclear matter territory.

A variety of new probes and results are coming out from the first ion runs at LHC. Even though the ALICE detector was the setup officially designed to study heavy ion collisions, both ATLAS and CMS are pursuing a rich program devoted to the study of the nuclear matter at extreme energy density, such as reached in heavy ion collisions. After the first two data taking periods with Lead ions, and an exploratory short pilot run with p-Pb collisions in 2012, a series of specific results originating from such studies has emerged and is currently discussed in the heavy ion community.

Understanding the bulk of reaction products in such collision events is mandatory to explore the role of specific processes observed in heavy ion collisions, so that it is generally accepted that, beside their interest in itself, proton-proton collision events are an essential reference for a comparison to data obtained in proton-nucleus or nucleus-nucleus collisions.

While the discussion of the large variety of physics results and experimental probes employed in such studies largely exceeds the scope of the present paper, the production of hadronic particles in proton-proton and heavy-ion collisions, which has now been studied by all large experiments running at CERN LHC, is the topic of interest for this contribution.

3. Experimental setups at the Large Hadron Colliders

Several large detectors have been built and are currently operating at the Large Hadron Collider. Some of them are specialized detectors, which aim at the investigation of specific aspects of the proton-proton and heavy ion collisions. General purpose detectors, which have the required performance to address all the main features of particle production in ultra-relativistic pp and AA collisions are the ALICE (A Large Ion Collision Experiment) [1], ATLAS (A Toroidal LHC Apparatus) [2] and CMS (Compact Muon Solenoid) [3] installations, while LHCb (LHCbeauty) [4] was designed especially for precision measurements of CP violation and rare decays of B hadrons, exploring the forward rapidity region.

Due to the experimental conditions at the LHC, all the detectors required fast, radiation-hard electronics and sensor elements. In addition, a high detector granularity was needed to handle the particle fluxes and to reduce the influence of overlapping events. For the study of the global properties of particle production large acceptance in pseudorapidity with almost full azimuthal angle coverage is required. Good charged-particle momentum resolution and reconstruction efficiency in the inner tracker are also essential.

While a full description of such detectors has been given, prior to the startup of LHC, in Ref. [1, 2, 3, 4], here only a brief recall of their main features is reported.

3.1. The ALICE detector

The choice and design of ALICE was driven by the physics requirements as well as by the experimental conditions expected in nucleus-nucleus collisions at the LHC. The most stringent design constraint is the extreme particle multiplicity. The design of ALICE was originally optimized for a value of about $dN_{ch}/d\eta = 4000$, and tested with simulations up to twice such value.

The overall dimensions of the ALICE detector are 16x16x26 m^3, with a total weight of approximately 10,000 tons. ALICE consists of a central barrel, which measures hadrons, electrons, and photons, and a forward muon spectrometer. The central part covers polar angles from 45° to 135° and is embedded in a large magnet, with a magnetic field of up to 0.5 T. The central barrel contains a set of tracking detectors and a set of particle identification detectors, followed by two different electromagnetic calorimeters for photon and jet measurements. Most of the central detectors cover the full azimuthal range. On one side, at small angles relative to the beam direction is a muon spectrometer which includes a dipole magnet generating a maximum field of 0.7 T. Additional detectors are placed at forward rapidity for event selection, timing reference and to measure global features of the reaction.

3.2. The ATLAS detector

The ATLAS detector has nominally a forward-backward symmetry with respect to the interaction point. The magnet configuration includes a thin superconducting solenoid surrounding the inner-detector, and three large superconducting toroids (one barrel and two end-caps) arranged with an eight-fold azimuthal symmetry around the calorimeters. The inner detector is immersed in a 2 T solenoidal field, has full coverage in ϕ and covers the pseudorapidity range $|\eta| < 2.5$.

Pattern recognition, momentum and vertex measurements, and electron identification are achieved with a combination of pixel and strip detectors in the inner part of the tracking volume, and straw-tube tracking detectors.

Extensive calorimetry, with different techniques, is also used in the ATLAS installation, to provide precision measurements of electrons and photons around the inner tracker and good reconstruction of jets and missing energy in the overall pseudorapidity range.

3.3. The CMS detector

The CMS detector is a general purpose installation able to cope with all the main signatures and probes of the high-luminosity pp collisions at LHC. The central feature of the CMS apparatus is a superconducting solenoid, 15 m long and 6 m internal diameter, with a 3.8 T field. In the field volume several detectors are located: the silicon pixel and strip tracker, the lead tungstate electromagnetic calorimeter, the brass/scintillator hadron calorimeter and the muon detection system. CMS has also extensive forward calorimetry. The tracker measures charged particles within the pseudorapidity range $-2.4 < \eta < 2.4$, providing an impact-parameter resolution of about 100 μm and an absolute p_T resolution of about 0.7 % for charged particles of momentum 1 GeV/c, of relevance for the study of particle production.

3.4. The LHCb detector

The LHCb experiment is dedicated to precision measurements of CP violation and rare decays of B hadrons. The detector is is a single-arm magnetic dipole spectrometer with a polar angular coverage with respect to the beam line of approximately 15 to 300 mrad in the horizontal bending plane, and 15 to 250 mrad in the vertical non-bending plane.

The LHCb spectrometer acceptance, $-2.5 < \eta < -2.0$ and $2.0 < \eta < 4.5$, allows also the forward pseudorapidity region to be probed. For the reconstruction of charged particles, the LHCb tracking system consists of silicon microstrip modules surrounding the proton-proton

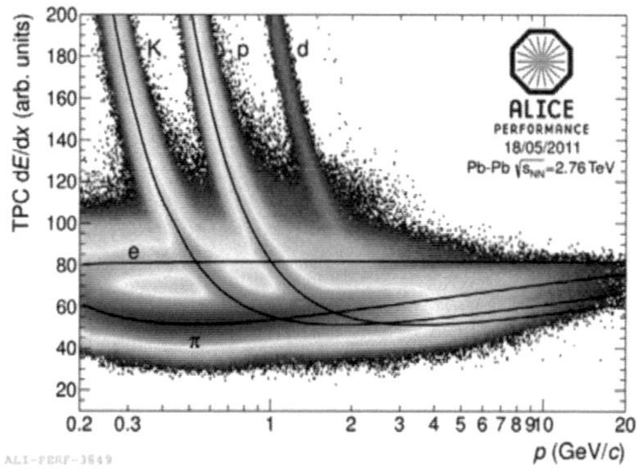

Figure 1. Particle identification performance in the ALICE TPC.

interaction region, a tracking station before the dipole magnet, which also employs microstrip technology, and three tracking stations - with silicon microstrips and straw tubes - after the magnet.

3.5. Particle reconstruction and identification

The study of particle production, especially in a large multiplicity heavy ion collision, requires optimal capabilities of the detector, in terms of primary and secondary vertex reconstruction, impact parameter and momentum resolution, tracking efficiency. As an example, in the ALICE detector, tracking in the central barrel is provided by an Inner Tracking System (ITS), a six-layer, silicon vertex detector, and by a large Time-Projection Chamber (TPC). A Transition Radiation Detector (TRD) is also used for tracking in the central region improving the transverse momentum resolution at high momentum. Because of the high particle density, the innermost four layers need to be truly two-dimensional devices, i.e. silicon pixel and silicon drift detectors, while the outer layers are equipped with double-sided silicon microstrip detectors. The need for efficient and robust tracking has led to the choice of a TPC as the main tracking detector. With its large granularity, this tracking device can guarantee reliable performance even in a large multiplicity environment.

Particle identification (PID) over a large part of the phase space and for many different particles is an important design feature of any detector at LHC.

In case of the ALICE detector, essentially all known PID techniques are employed: specific ionization energy loss dE/dx, time-of-flight, transition and Cherenkov radiation, electromagnetic calorimetry, muon filters, and topological decay reconstruction. As an example, Fig.1 shows the typical particle identification performance plot for the ALICE TPC.

Even though all the LHC detectors have wide capabilities for the reconstruction and identification of primary and secondary particles emitted in the collisions, several differences exist between them, in terms of geometrical acceptance, impact parameter and momentum resolution, PID efficiency, particle detection thresholds, and so on. For a detailed understanding of the capabilities of each LHC detector the reader is addressed to their general description [1, 2, 3, 4].

4. Results from proton-proton collisions

4.1. Multiplicity

The multiplicity of charged particles produced in a high energy collision is a key quantity to characterize the hadronic final state. The pseudorapidity density and the multiplicity distributions for primary charged particles are among the basic observables to be measured

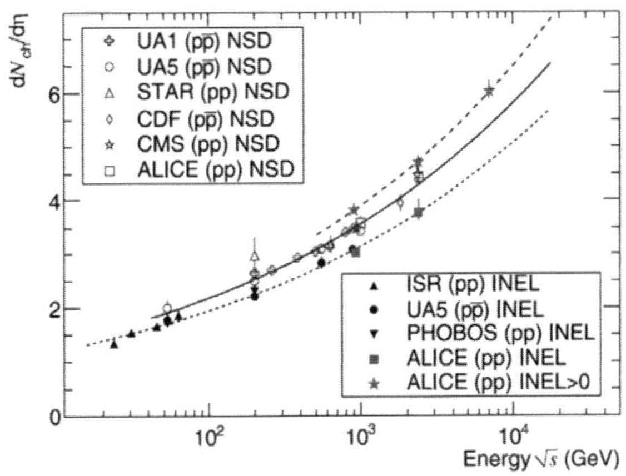

Figure 2. Pseudorapidity density of charged particles measured at LHC by different experiments in the central pseudorapidity density [5].

and understood. Several predictions from theoretical models were refined in recent years, taking into account data from RHIC, so that the first data from LHC were of crucial importance to check the ability of such models to reproduce new data at the higher energy regime. These observables were indeed measured at LHC by the various Collaborations as soon as a new beam energy was available for physics.

ALICE, ATLAS and CMS have measured charged particle multiplicities in the central pseudorapidity region, while LHCb has mainly covered the forward region.

Concerning the evolution of the average charged particle multiplicity with the centre-of-mass energy, the ALICE Collaboration has reported values of the charged particle pseudorapidity densities $dN_{ch}/d\eta$, measured in the central pseudorapidity region $|\eta| < 1$, equal to 3.81 ± 0.01, 4.70 ± 0.01 and 6.01 ± 0.01 (errors are statistical only) at 0.9, 2.36 and 7 TeV centre-of-mass energy respectively [5]. It was observed that the experimental values of such pseudorapidity densities are generally higher with respect to model predictions, except in a few cases. The data obtained at 7 TeV demonstrate, in agreement with the result from CMS [6], that the measured multiplicity density increases with the energy much more than provided by the models considered.

The energy dependence of the charged-particle pseudorapidity density in the central pseudorapidity region has been extracted from the various data sets measured at LHC and compared for different event classes, see Fig.2 [5]. The lines indicate a power law fit to the data.

4.2. Multiplicity distributions

Multiplicity distributions were measured for various ranges of the pseudorapidity, both in the central pseudorapidity region and at forward rapidities. Typical results are shown in fig.3, as reported by the CMS Collaboration [7], at 0.9, 2.36 and 7 TeV centre-of-mass energy, compared to earlier measurements performed by UA5 and to the results from the ALICE Collaboration [5]. Similar set of results in the pseudorapidity interval $|\eta| < 2.5$ have been also reported by the ATLAS Collaboration [8].

Figure 4 shows unfolded charged particle multiplicity distribution for different bins in pseudorapidity, as measured by the LHCb Collaboration [9], together with the predictions from different event generators. None of them is fully able to describe these multiplicity distributions over the full LHCb acceptance. In general, the models underestimate charged

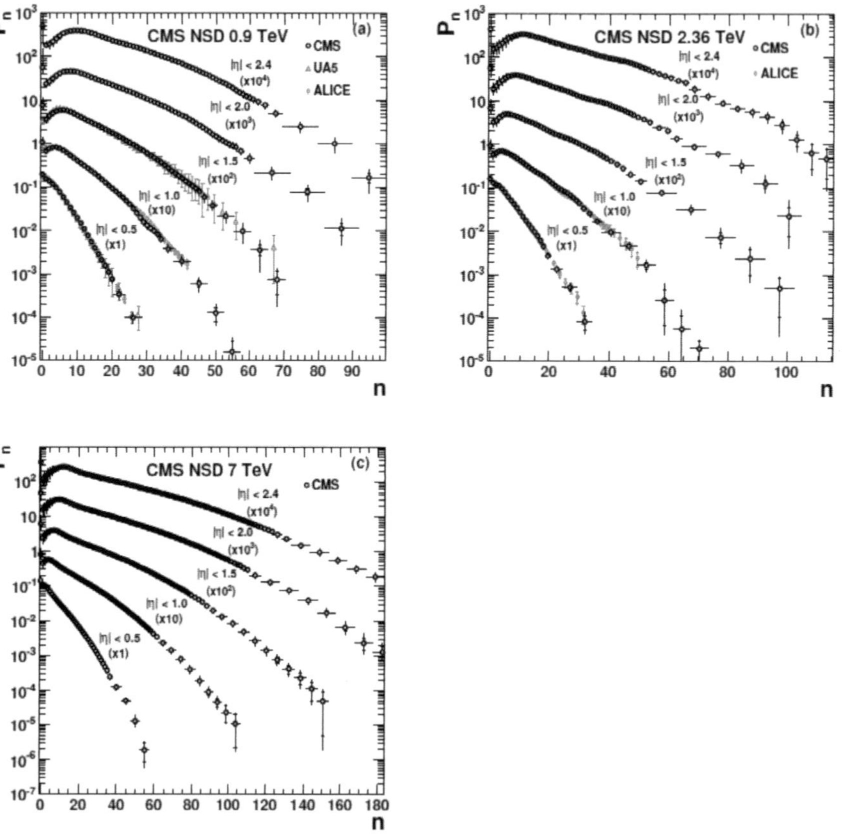

Figure 3. Charged hadron multiplicity spectrum, measured at 0.9, 2.36 and 7 TeV in increasing ranges of the pseudorapidity, from 0.5 to 2.4. Results from the CMS and ALICE Collaborations are shown, togeher with the results of earlier measurements from the UA5 Collaboration [7].

particle production, in agreement with the measurements by the other LHC experiments in the central region.

4.3. Pseudorapidity distributions

The charged particle pseudorapidity density as a function of the pseudorapidity has been measured at various energies, both in the central region, and at forward rapidity by LHCb [9], and compared with the current predictions of event generators.

For instance, high-statistics measurements of the charged primary particle pseudorapidity density in pp collisions at centre-of-mass energies of 0.9 TeV and 2.36 TeV with the ALICE detector have been reported in Ref. [10]. The results at 0.9 TeV are consistent with UA5 earlier measurements of antiproton-proton at the same energy, and consistent at both energies with the CMS measurements. A comparison with various models shows that none of the investigated models and tunes describes the results well. In particular, they underestimate the increase in the average multiplicity seen in the data between 0.9 TeV and 2.36 TeV.

The ATLAS Collaboration has measured charged-particles pseudorapidity distributions at centre-of-mass energies of 0.9, 2.36 and 7 TeV under different trigger conditions. As an example, fig.5 shows the result at 7 TeV in the range $|\eta| < 2.5$, under the condition $n_{ch} > 1$ and $p_T > 500$

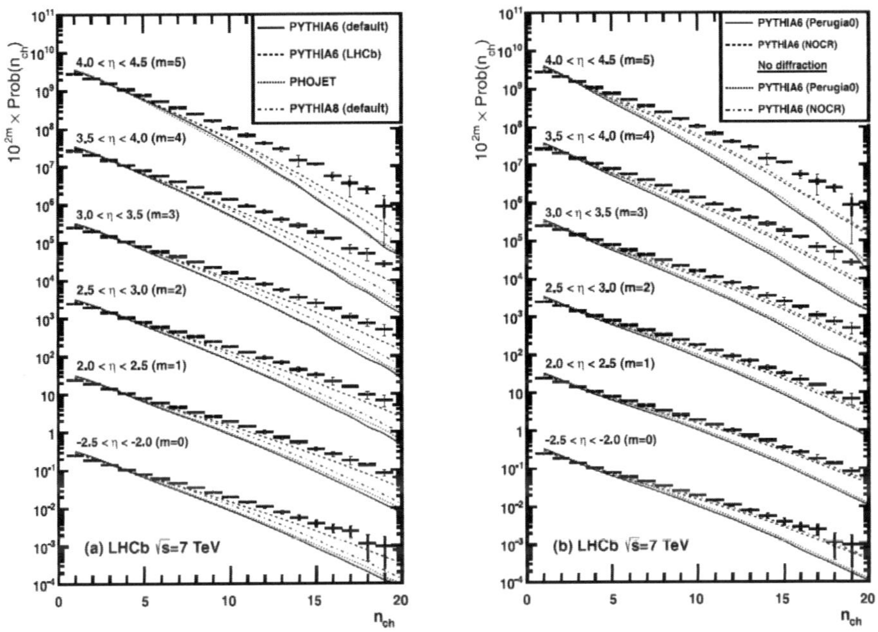

Figure 4. Charged particle multiplicity distributions for different bins in pseudorapidity, as reported by the LHCb Collaboration [9].

MeV, together with the predictions from various models.

In the forward region, $\eta > 2.5$, minimum bias events measured by LHCb were seen to give a distribution not reproduced by any of the models considered. Different event selections were also considered to extract hard QCD events from the data sample, for instance by imposing that at least one track in the forward region has transverse momentum larger than 1 GeV/c. Although the predictions are in such case in better agreement than for minimum bias events, all models still fail to provide a detailed description of the average charged particle multiplicity per unit of pseudorapidity [9].

4.4. Transverse momentum spectra

Transverse momentum spectra and yields of identified hadrons at midrapidity are the next key quantities to be measured in high energy collisions. Such data allow to extract the detailed shape of the transverse momentum spectra, the average transverse momentum, the ratios between the yields of the various species and the antiparticle/particle ratios, thus providing a good set of benchmarks for the currently employed theoretical models.

A first analysis of identified hadrons (π^+, π^-, K^+, K^-, p, $\bar{p}$) at 0.9 TeV has been carried out by ALICE, down to very low p_T, see Fig.6 [11]. A fit of the data with a Levy function was seen to reproduce the data well and was used to extract the average p_T and the total yields of the various species. Transverse momentum spectra of charged positive and negative hadrons were published by CMS at 0.9 and 7 TeV [12] in a very large momentum range, up to 200 GeV/c and in the pseudorapidity interval $|\eta| < 2.4$. Fig.7 shows this result, compared to the predictions of different PYTHIA tunes.

Spectra of identified hadrons produced in pp collisions at 0.9, 2.36 and 7 TeV have also been measured at midrapidity by CMS in the transverse momentum region 0.1-1.7 GeV/c and

Journal of Physics: Conference Series **424** (2013) 012004 doi:10.1088/1742-6596/424/1/012004

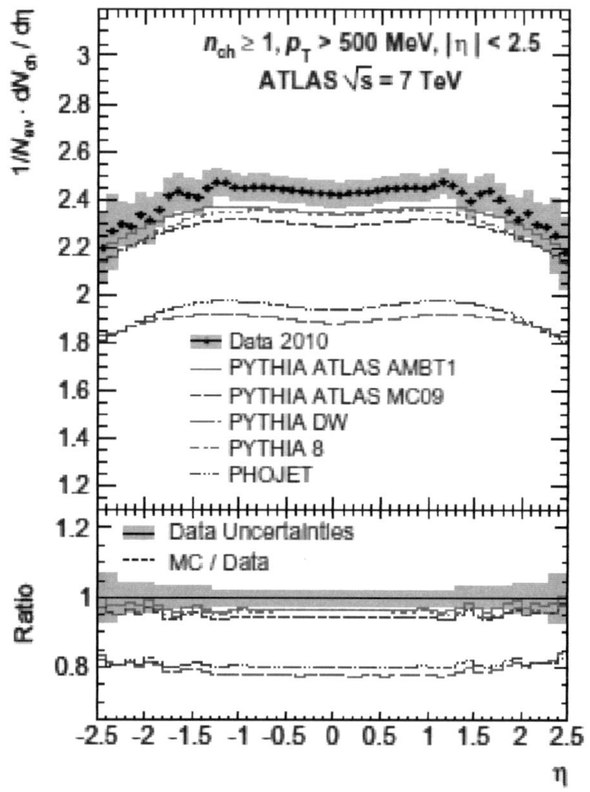

Figure 5. Inclusive charged-particle multiplicity as a function of the pseudorapidity, in the interval $|\eta| < 2.5$ and p_T >500 MeV, measured by the ATLAS Collaboration [8].

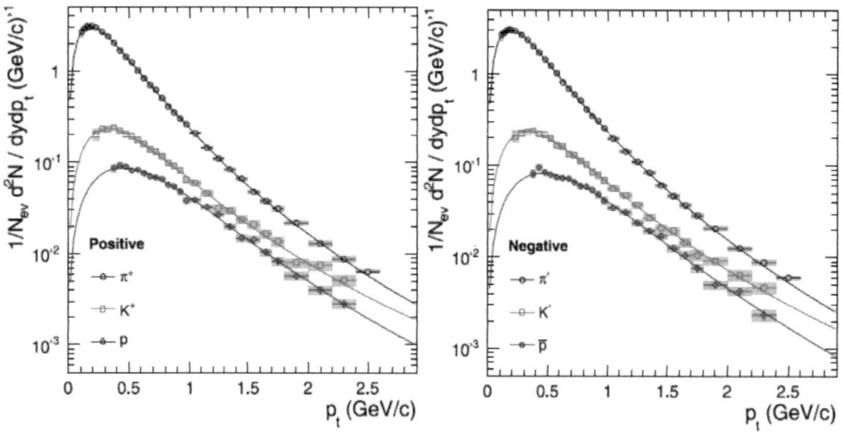

Figure 6. Transverse momentum spectra of identified hadrons measured at 0.9 TeV by the ALICE Collaboration [11].

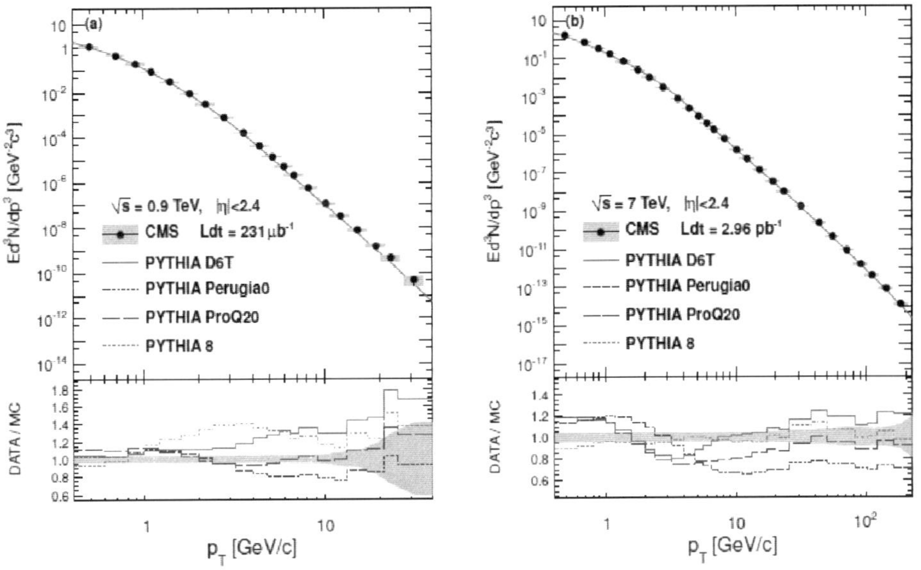

Figure 7. Transverse momentum spectra of charged hadrons measured at 0.9 and 7 TeV by the CMS Collaboration in a large momentum range [12].

$|y| < 1$ [13]. Fig.8 shows the experimental distributions obtained for pions, kaons and protons, together with the corresponding negative hadrons. A selection of events according to their multiplicity has also been made, allowing to study particle production as a function of the event multiplicity. The results show that particle production has a strong correlation with such variable, rather than with the centre-of-mass energy [13].

The average transverse momentum as a function of the particle multiplicity, as reported by the ATLAS Collaboration [8] is shown in fig.9, at the centre-of-mass energies of 0.9 and 7 TeV, together with its ratio to the value provided by different models.

5. Heavy ion collisions

Heavy ion collisions at LHC were believed to generate a system at unprecedented temperatures and energy densities, thus opening a new frontier in the study of QCD matter. The basic step in order to characterize this system is the measurement of the charged particle pseudorapidity density, which is an essential observable to estimate the initial energy density and strongly constrains the various proposed production mechanisms. They predicted values differing by a factor 2 before the new LHC heavy ion data could be obtained.

The first measurement of the charged-particle pseudorapidity density produced in central Pb-Pb collisions at a centre-of-mass energy per nucleon pair $\sqrt{s_{NN}} = 2.76$ TeV was obtained with the ALICE detection setup [14]. Primary charged particles, including decay products (except those originating from the weak decay of strange particles) were measured in the pseudorapidity interval $|\eta| < 0.5$, thus obtaining the value of the primary charged-particle density $dN_{ch}/d\eta$ in central Pb-Pb collisions. A value of $dN_{ch}/d\eta = 1584 \pm 4$ (stat) $\pm$ 76 (syst) was obtained, which corresponds to a normalized value per participant pair of $dN_{ch}/d\eta/(0.5 < N_{part} >) = 8.3 \pm 0.4$ (syst).

The evolution of the charged-particle pseudorapidity density per participant pair with the centre-of-mass energy is shown in Fig.10, for central nucleus-nucleus collisions and for non single

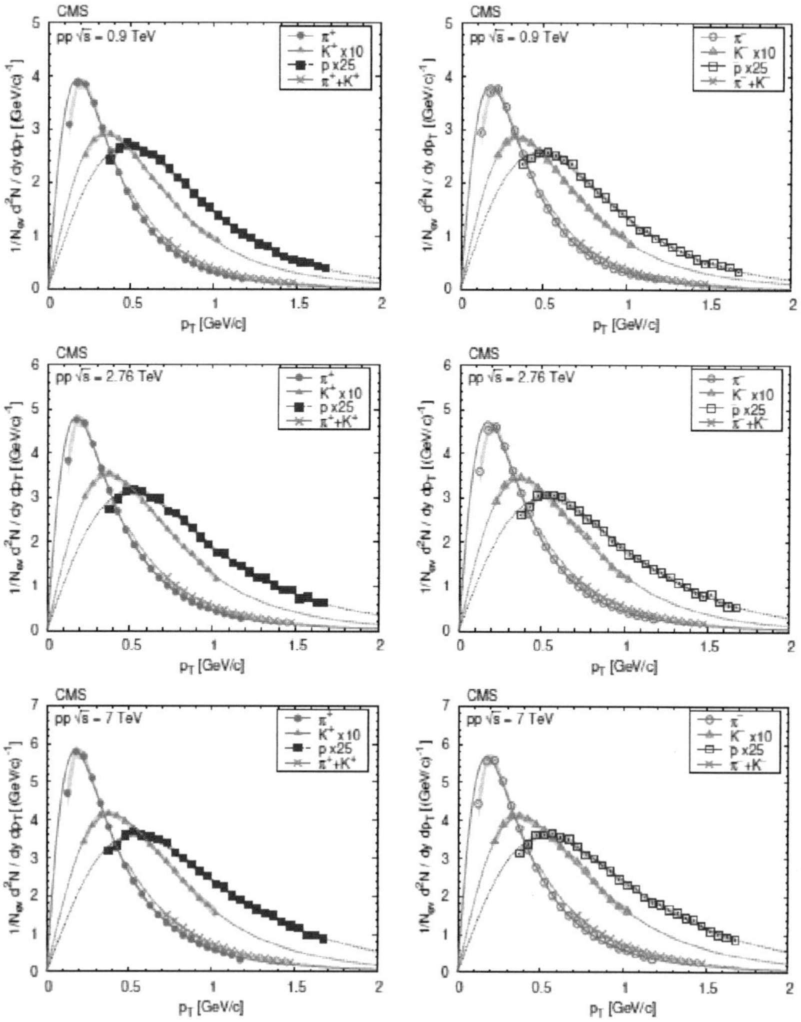

Figure 8. Transverse momentum spectra of identified hadrons measured at 0.9, 2.36 and 7 TeV by the CMS Collaboration [13].

diffractive pp collisions. Solid lines in this plot show the curves $(s_{NN})^n$, where the exponent n is larger (0.15) for nucleus-nucleus and smaller (0.11) for pp collisions, demonstrating a steeper energy dependence for heavy ion collisions. Comparing this value to that obtained for pp and $p\bar{p}$ collisions at similar energies, an increase by a factor 1.9 is observed.

An increase of a factor 2.2 is instead observed when comparing such value to those previously measured at RHIC for Au-Au collisions at a centre-of-mass energy $\sqrt{s}_{NN} = 0.2$ TeV.

The observed value of this quantity was compared to the various predictions which were available prior to obtain the LHC data, following their tuning to RHIC data at 0.2 TeV (fig.11). Some of them, based on the HIJING model tuned to pp data at 7 TeV, on the dual parton model or on the ultra-relativistic quantum molecular dynamics model, are consistent with the measurement. Other models exhibit various levels of agreement, with several of them sensibly underestimating the experimental result, and a hydrodynamical model, with a scaling of the

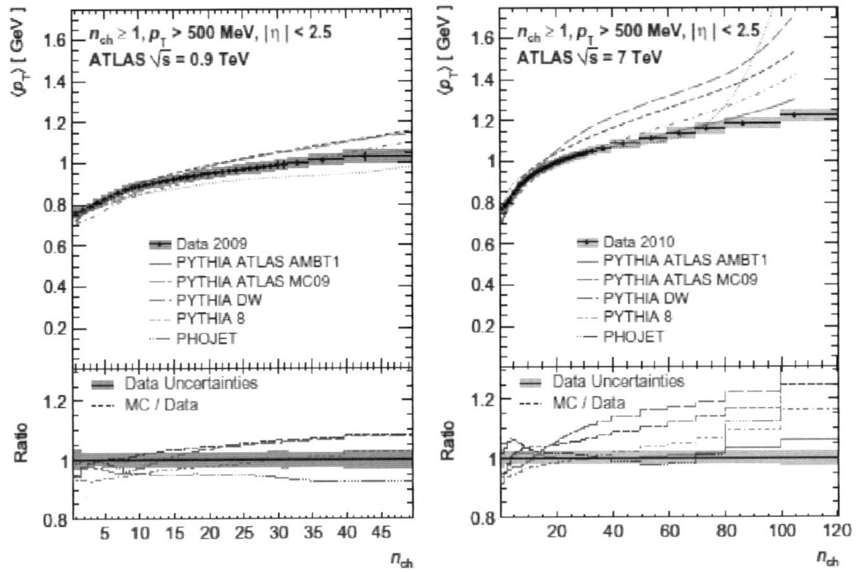

Figure 9. Average transverse momentum as a function of the charged particle multiplicity, as reported by the ATLAS Collaboration [8].

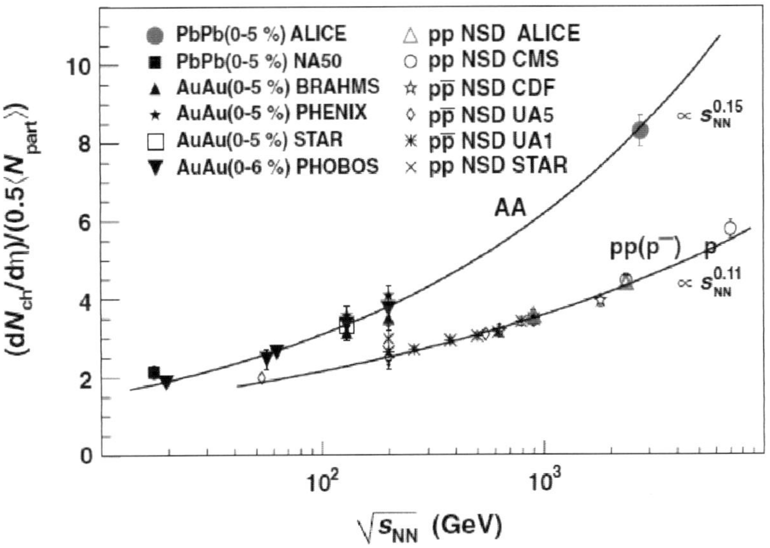

Figure 10. Charged-particle pseudorapidity density per participant pair as a function of $\sqrt{s_{NN}}$, for central nucleus-nucleus collisions and nonsingle diffractive pp and $p\bar{p}$ collisions [14].

multiplicity from pp collisions, strongly overestimating the result.

Heavy ion collisions may be characterized by centrality, thus allowing to extend the understanding of the contribution of hard and soft processes to particle production. It is thus important to measure the dependence of the charged particle multiplicity upon the centrality.

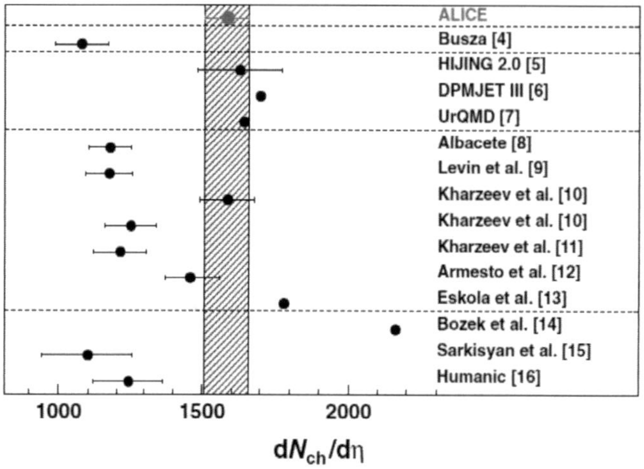

Figure 11. Comparison of the experimental value of the charged-particle pseudorapidity density, as obtained by the ALICE Collaboration [14], to the existing theoretical predictions available at the time.

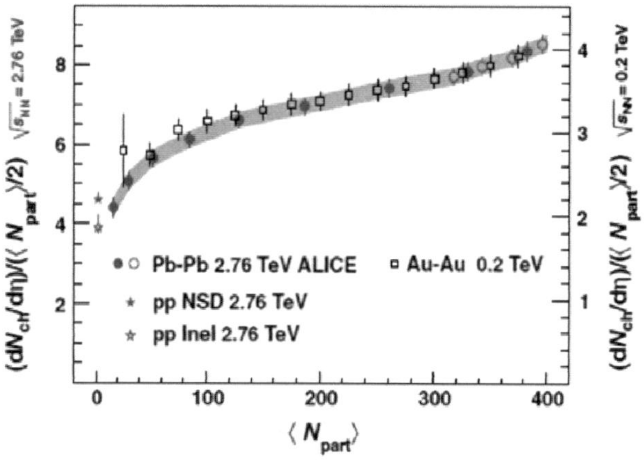

Figure 12. Dependence of $(dN_{ch}/d\eta)/(< N_{part} > /2)$ on the number of participants, for Pb-Pb collisions at $\sqrt{s}_{NN} = 2.76$ TeV (ALICE Collaboration [15]), compared to Au-Au data at $\sqrt{s}_{NN}$ = 0.2 TeV from RHIC.

This has been done by the ALICE Collaboration [15] by appropriate selection of the collision events into nine centrality classes, covering the most central 80 % of the hadronic cross section. As it is seen in Fig.12, the charged-particle density per participating nucleon pair increases by a factor 2 when going from peripheral to central collisions, and its trend is similar to that observed at lower centre-of-mass energy. A comparison to different theoretical predictions shows that some models (HIJING 2.0, or different versions of saturation models), which were tuned after the most central $(dN_{ch}/d\eta)$ value was published, reasonably reproduce the trend.

6. Conclusion

Systematic measurements of the main features of particle production in the new energy regime of the Large Hadron Collider have been undertaken by all major LHC Collaborations, for proton-proton and heavy ion collisions. Charged-particle multiplicities have been extracted from the analysis of collision events at different centre-of-mass energies, and in different pseudorapidity regions. Multiplicity distributions have been reported, with a high statistics, such as to observe high-multiplicity events even in pp collisions at midrapidity. Forward rapidity distributions have been measured as well, up to η=4.5. Moreover, the pseudorapidity distributions have been measured in a wide pseudorapidity range and under different trigger conditions. Transverse momentum spectra of charged (positive and negative) hadrons have been measured as well, in a huge momentum range, and the average transverse momentum extracted as a function of the particle multiplicity. Concerning heavy ion collisions, the very first Pb-Pb data have made it possible to measure the charged particle multiplicity, which was one of the expected new results at LHC, allowing to characterize the collision dynamics. Also differential results, as a function of the centrality, have been already obtained.

As a result of these new experimental findings, a lot of efforts has been put to provide a systematic comparison to model predictions, either tuned on the old RHIC results or on the basis of the first LHC pp data, and this has given new inputs for an improvement of the theoretical descriptions.

Work is still in progress however on the experimental aspects of the understanding of the detailed features of particle production at LHC. Additional data analysis is in progress by all the Collaborations to extend the systematics to identified hadrons and to other particle species, such as those particles which are reconstructed through their decay. The pilot run on pPb collisions, which will be extended in 2013, is believed to provide a lot of new data helpful to understand the transition between pp and nucleus-nucleus collisions. A corresponding effort will be put on the improvement of the existing theoretical descriptions of the collision dynamics. This will help to understand the particle production mechanisms at high energy in the overall context of the properties of matter under extreme conditions.

7. References

[1] The ALICE Collaboration, JINST 3(2008)S08002.
[2] The ATLAS Collaboration, JINST 3(2008)S08003.
[3] The CMS Collaboration, JINST 3(2008)S8004.
[4] The LHCb Collaboration, JINST 3(2008)S08005.
[5] K.Aamodt et al. (The ALICE Collaboration), Eur. Phys. J. C68(2010)345.
[6] V.Khachatryan et al. (The CMS Collaboration), JHEP02(2010)041.
[7] S.Chatrchyan et al. (The CMS Collaboration), JHEP01(2011)079.
[8] G.Aad et al. (The ATLAS Collaboration), New J. Phys. 13(2011)053033.
[9] R.Aaij et al. (The LHCb Collaboration), Eur. Phys. J. C72(2012)1947.
[10] K.Aamodt et al. (The ALICE Collaboration), Eur. Phys J C68(2010)89.
[11] K.Aamodt et al. (The ALICE Collaboration), Eur. Phys. J. C71(2011)1655.
[12] S.Chatrchyan et al. (The CMS Collaboration), JHEP 08(2011)086.
[13] S.Chatrchyan et al. (The CMS Collaboration), Eur. Phys. J. C.72(2012)2164.
[14] K.Aamodt et al. (The ALICE Collaboration), Phys. Rev.Letters 105(2010)252301.
[15] K.Aamodt et al. (The ALICE Collaboration), Phys. Rev. Letters 106(2011)032301

Journal of Physics: Conference Series **424** (2013) 012005 doi:10.1088/1742-6596/424/1/012005

Kaon interferometry at KLOE/KLOE–2[1]

Eryk Czerwiński

Institute of Physics, Jagiellonian University, ul. Reymonta 4, 30-059 Cracow, Poland

E-mail: eryk.czerwinski@uj.edu.pl

Abstract. Neutral kaons from ϕ meson decays are produced in pairs, which offers the possibility to select pure kaon beams: the detection of a K_S (K_L) tags the presence of a K_L (K_S), the same holds for charged kaons. This allows to perform precise measurement of kaon properties. Additionally, these neutral kaon pairs are produced in a pure quantum state ($J^{PC} = 1^{--}$), which allows to investigate CP and CPT symmetries via quantum interference effects, as well as the basic principles of quantum mechanics.

Introduction to kaon interferometry and the results on the decoherence parameter and CPT test obtained by the KLOE experiment at the DAΦNE e^+e^- collider will be presented together with prospects for KLOE–2 project.

1. Introduction

The initial state of the two neutral kaons originated from ϕ meson ($J^{PC} = 1^{--}$) has to be antisymmetric and can be expressed as [1]:

$$|i\rangle = \frac{1}{\sqrt{2}} \left[\left|K^0(+\vec{p})\right\rangle \left|\bar{K}^0(-\vec{p})\right\rangle - \left|\bar{K}^0(+\vec{p})\right\rangle \left|K^0(-\vec{p})\right\rangle \right] \tag{1}$$

with the requirement: $C|i\rangle = -|i\rangle$ and $P|i\rangle = -|i\rangle$. For the description of kaons production the strangeness basis can be used $\{\left|K^0\right\rangle, \left|\bar{K}^0\right\rangle\}$, while for the decay studies it can be changed to the $\{|K_S\rangle, |K_L\rangle\}$ one:

$$|i\rangle = \frac{N}{\sqrt{2}} \left[|K_S(+\vec{p})\rangle |K_L(-\vec{p})\rangle - |K_L(+\vec{p})\rangle |K_S(-\vec{p})\rangle \right], \tag{2}$$

where N is a normalization factor equal to:

$$N = \frac{\sqrt{(1 + |\epsilon_S|^2)(1 + |\epsilon_L|^2)}}{1 - \epsilon_S \epsilon_L} \approx 1. \tag{3}$$

[1] On behalf of KLOE/KLOE–2 collaborations: D. Babusci, D. Badoni, I. Balwierz-Pytko, G. Bencivenni, C. Bini, C. Bloise, F. Bossi, P. Branchini, A. Budano, L. Caldeira Balkeståhl, G. Capon, F. Ceradini, P. Ciambrone, F. Curciarello, E. Czerwiński, E. Danè, V. De Leo, E. De Lucia, G. De Robertis, A. De Santis, A. Di Domenico, C. Di Donato, R. Di Salvo, D. Domenici, O. Erriquez, G. Fanizzi, A. Fantini, G. Felici, S. Fiore, P. Franzini, P. Gauzzi, G. Giardina, S. Giovannella, F. Gonnella, E. Graziani, F. Happacher, L. Heijkenskjöld B. Höistad, L. Iafolla, M. Jacewicz, T. Johansson, A. Kupsc, J. Lee-Franzini, B. Leverington, F. Loddo, S. Loffredo, G. Mandaglio, M. Martemianov, M. Martini, M. Mascolo, R. Messi, S. Miscetti, G. Morello, D. Moricciani, P. Moskal, F. Nguyen, A. Passeri, V. Patera, I. Prado Longhi, A. Ranieri, C. F. Redmer, P. Santangelo, I. Sarra, M. Schioppa, B. Sciascia, M. Silarski, C. Taccini, L. Tortora, G. Venanzoni, W. Wiślicki, M. Wolke, J. Zdebik

Since short- and long-lived states evolve in time as pure exponentials, the decay amplitude of such system can be presented as:

$$A(f_1,t_1;f_2,t_2) = \frac{N}{\sqrt{2}}\left[\langle f_1|T|K_S(t_1)\rangle \langle f_2|T|K_L(t_2)\rangle - \langle f_1|T|K_L(t_1)\rangle \langle f_2|T|K_S(t_2)\rangle\right] =$$
$$= \frac{N}{\sqrt{2}}\left[\langle f_1|T|K_S\rangle \langle f_2|T|K_L\rangle e^{-i\lambda_S t_1}e^{-i\lambda_L t_2} - \langle f_1|T|K_L\rangle \langle f_2|T|K_S\rangle e^{-i\lambda_L t_1}e^{-i\lambda_S t_2}\right], \quad (4)$$

where f_1 and f_2 denote decay final states at kaon proper times t_1 and t_2, T is the transition matrix and the eigenvalues:

$$\lambda_S = m_S - i\frac{\Gamma_S}{2} \quad \text{and} \quad \lambda_L = m_L - i\frac{\Gamma_L}{2}. \quad (5)$$

The double decay rate for $\phi \to K^0\bar{K}^0 \to f_1 f_2$ can be computed from equation (4):

$$\begin{aligned} I(f_1,t_1;f_2,t_2) &= |A(f_1,t_1;f_2,t_2)|^2 = C_{12}\Big[|\eta_1|^2 e^{-\Gamma_L t_1 - \Gamma_S t_2} + |\eta_2|^2 e^{-\Gamma_S t_1 - \Gamma_L t_2} + \\ &- 2|\eta_1||\eta_2| e^{-\frac{(\Gamma_S+\Gamma_L)}{2}(t_1+t_2)} \cos(\Delta m(t_1-t_2) + \varphi_2 - \varphi_1)\Big], \end{aligned} \quad (6)$$

where φ_1 and φ_2 are phases and:

$$\begin{aligned} C_{12} &= \frac{|N|^2}{2}\left|\langle f_1|T|K_S\rangle \langle f_2|T|K_S\rangle\right|^2, \\ \eta_i &= |\eta_i| e^{i\varphi_i} \equiv \frac{\langle f_i|T|K_L\rangle}{\langle f_i|T|K_S\rangle} \end{aligned} \quad (7)$$

and

$$\Delta m = m_L - m_S > 0 \quad \text{and} \quad \Delta\Gamma = \Gamma_S - \Gamma_L > 0. \quad (8)$$

For comparison of experimental data and above-mentioned theoretical evaluations the integration of the formula (6) over $t = t_1 + t_2$ at a fixed difference of time $\Delta t = t_1 - t_2$ can be done [1]:

$$\begin{aligned} I(f_1,f_2;\Delta t \geq 0) &= \frac{C_{12}}{\Gamma_S+\Gamma_L}\Big[|\eta_1|^2 e^{-\Gamma_L \Delta t} + |\eta_2|^2 e^{-\Gamma_S \Delta t} + \\ &- 2|\eta_1||\eta_2| e^{-\frac{(\Gamma_S+\Gamma_L)}{2}\Delta t}\cos(\Delta m \Delta t + \varphi_2 - \varphi_1)\Big], \end{aligned} \quad (9)$$

valid for $\Delta t \geq 0$, while for $\Delta t < 0$ the substitutions $\Delta t \to |\Delta t|$ and $1 \leftrightarrow 2$ have to be applied.

Equation (9) implies that both kaons cannot decay at the same time in the same final state $f_1 = f_2$, and it is an example of correlation of the type pointed out for the first time by Einstein, Podolsky and Rosen [2] and the last line of this equation shows a time interference term which is a characteristic correlation between both kaon decays [1].

2. DAΦNE collider and KLOE experiment

The e^+e^- collider and ϕ-factory DAΦNE is located in National Laboratory in Frascati (LNF-INFN, Italy). Electrons and positrons are accelerated in linac, then stored and cooled in an accumulator and finally transferred in bunches into separated storage rings with two interaction points [3]. A schematic view of DAΦNE complex is presented in Fig. 1. The collider was designed to operate at the peak of ϕ resonance ($\sqrt{s} = m_\phi \approx 1019\ MeV$). Since electrons and positrons

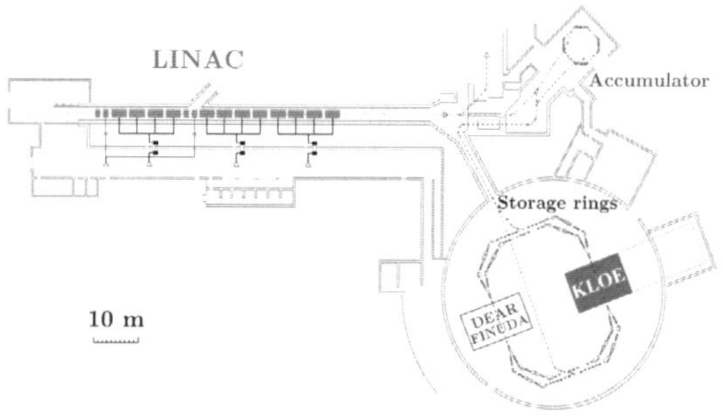

Figure 1. Scheme of the DAΦNE complex. A position of the KLOE detector at one of two interaction points is also presented. Figure taken from [4].

collide with small transverse momenta, the produced ϕ mesons are almost at rest ($\beta_\phi \approx 0.015$). A ϕ meson decays mostly into kaon pairs (49% into K^+K^- and 34% into K_SK_L), which makes a ϕ-factory the natural place for kaon physics studies.

In years 2001-2006 KLOE has collected 2.5 fb^{-1} of integrated luminosity, which corresponds to about 6.6×10^9 kaon pairs [5]. The detector itself consists of two main components: a ~3.3 m long cylindrical drift chamber [6] with radius ~4 m surrounded by an electromagnetic calorimeter [7]. Both sub-detectors are inserted in a superconducting coil which produces an axial magnetic field of 0.52 T parallel to the beam axis. Fig. 2 shows a schematic view of KLOE detector. The KLOE drift chamber was designed to detect a sizable fraction of K_L decays (mean

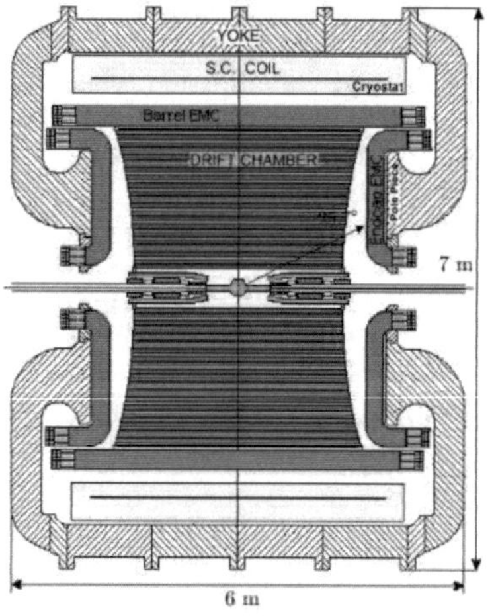

Figure 2. The KLOE detector surrounded by superconducting coil. Collision point of electrons and positrons is in the spherical beam-pipe in the center of the detector. Figure adapted from [4].

decay path of K_L meson is ~3.4 m). The chamber is filled with a mixture of helium and isobutane (90% and 10%, respectively) and has about 52000 wires arranged in cylindrical layers with alternate stereo angles. Based on the reconstruction of charged track curvature it allows for a fractional momentum accuracy of $\sigma_p/p \approx 0.5\%$. The resolution of vertex reconstruction is about 1 mm, while overall spatial accuracy is below 2 mm. An electromagnetic calorimeter consists of a barrel and side detectors (*endcaps*) providing almost 4π coverage of solid angle. Each module of calorimeter is read out on both sides by set of photomultipliers. It is build of stack of 1 mm scintillating fiber layers glued between grooved lead foils. The obtained accuracy of energy and time measurement are $\sigma_E = 5.7\%/\sqrt{E[GeV]}$ and $\sigma(t) = 54ps/\sqrt{E[GeV]} \oplus 100ps$, respectively. Determination of the hit position in the plane transverse to the fiber direction is based on the analysis of signal amplitude distribution and the resolution is about 1 cm, while accuracy of longitudinal coordinate due to excellent time resolution amounts to $\sigma_z = 1.2cm/\sqrt{E[GeV]}$.

Since at KLOE kaons are produced in pairs from ϕ decay, reconstruction of K_S decay close to interaction region (clean selection of $K_S \to \pi^+\pi^-$, BR=69%) allows for tagging of K_L presence. Having that and taking into account size of the detector itself makes KLOE an excellent place for K_L decay measurements. However, what is even more important, detection of K_L hit in calorimeter module tags the presence of K_S. This method makes KLOE an unique place to study pure K_S beams. This method was recently used for the most precise determination of the upper limit of $BR(K_S \to \pi^0\pi^0\pi^0)$ [8].

3. Decoherence parameter

The transition of a pure state into an incoherent mixture of states is generally denoted as decoherence [1], meaning that entanglement of particles is lost. The decoherence parameter ζ can be introduced by multiplying the interference term in equation (6) by a factor $(1-\zeta)$:

$$\begin{aligned} I(f_1,t_1;f_2,t_2) &= C_{12}\Big[|\eta_1|^2 e^{-\Gamma_L t_1 - \Gamma_S t_2} + |\eta_2|^2 e^{-\Gamma_S t_1 - \Gamma_L t_2} + \\ &- 2(1-\zeta)|\eta_1||\eta_2| e^{-\frac{(\Gamma_S+\Gamma_L)}{2}(t_1+t_2)} \cos(\Delta m(t_1-t_2) + \varphi_2 - \varphi_1)\Big]. \end{aligned} \tag{10}$$

Figure 3 shows sensitivity of the double decay rate distribution to the value of ζ in the case $f_1 = f_2 = \pi^+\pi^-$. The biggest discrepancy can be observed for Δt close to 0. The quantum mechanics case is described without decoherence ($\zeta = 0$), while the total decoherence corresponds to $\zeta = 1$, so kaons are no longer entangled. Intermediate situations between these two correspond to values of ζ within the range (0,1).

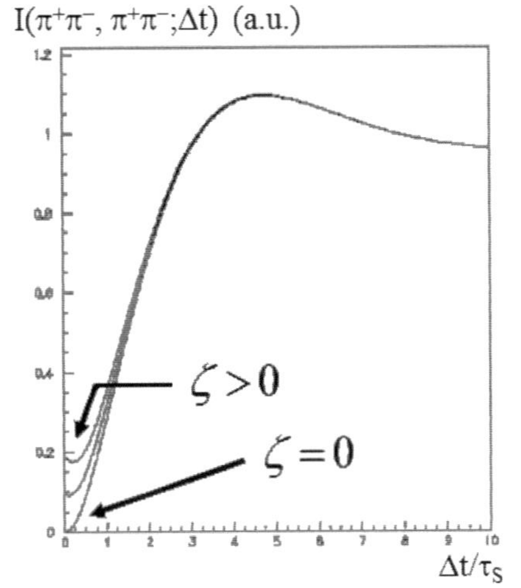

Figure 3. The $I(\pi^+\pi^-, \pi^+\pi^-, \Delta t)$ distribution for quantum mechanic case ($\zeta = 0$), and for two greater than zero values of the decoherence parameter. The biggest discrepancy is for Δt close to 0. The figure is taken from [1]

Journal of Physics: Conference Series **424** (2013) 012005 doi:10.1088/1742-6596/424/1/012005

Based on the integrated luminosity of ≈1.5 fb^{-1} study of the decoherence parameter was performed at KLOE [9]. The signal events $\phi \to K_S K_L \to \pi^+\pi^-\pi^+\pi^-$ were selected by requirement of two vertexes, each with two opposite-curvature tracks inside the drift chamber, with an invariant mass and total momentum compatible with the two neutral kaon decays. The experimental points were fitted with Eq. 11 modified with parameters expressing decoherence in different models described in [1,9]. The fit was performed taking into account resolution and detection efficiency, the background from coherent and incoherent K_S regeneration on the beam pipe wall, and the small contamination from the non-resonant $e^+e^- \to \pi^+\pi^-\pi^+\pi^-$ channel. The determined experimental distribution of the $\phi \to K_S K_L \to \pi^+\pi^-\pi^+\pi^-$ intensity as a function of the absolute value of Δt is shown in Fig. 4. The results:

$$\zeta_{SL} = (0.3 \pm 1.8_{stat} \pm 0.6_{syst}) \cdot 10^{-2} ,$$
$$\zeta_{0\bar{0}} = (1.4 \pm 9.5_{stat} \pm 3.8_{syst}) \cdot 10^{-7} , \tag{11}$$

where ζ_{SL} and $\zeta_{0\bar{0}}$ depend on the basis in which the initial state is expressed: $\{|K^0\rangle, |\bar{K}^0\rangle\}$ or $\{|K_S\rangle, |K_L\rangle\}$, show no deviation from quantum mechanics [9].

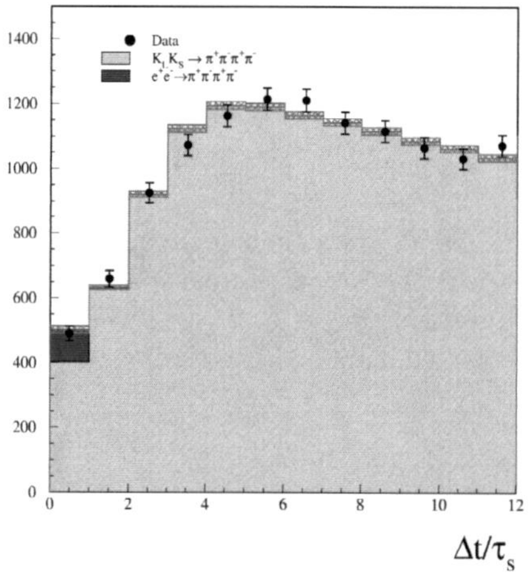

Figure 4. Points denote experimental results, while fitted histogram shows results of the Monte Carlo simulation. The bin size corresponds to the time resolution $\sigma(\Delta t) \approx \tau_S$ [9].

4. CPT, QM and Lorentz symmetry breaking

CPT invariance is a fundamental theorem in the framework of quantum field theory (QFT), as a consequence of Lorenz invariance, unitarity and locality. In several quantum gravity (QG) models, however, *CPT* can be violated via some mechanism which can also violate standard Quantum Mechanics (QM). In a review Bernabeu, Ellis, Mavromatos, Nanopoulos and Papavassiliou [10] discuss the theoretical motivations for possible *CPT* violations and the unique role played by the entangled neutral kaon pairs produced at DAΦNE in precision tests of the *CPT* symmetry.

As an example of this incredible precision reachable with neutral kaons, the model by Ellis, Hagelin, Nanopoulos and Srednicki (EHNS) which introduces three *CPT* and QM-violating real parameters α, β and γ [11] can be used. On phenomenological grounds, they are expected to be $O(m_K^2/M_{Pl}) \sim 2\times10^{-20}$ GeV at most, since $M_{Pl} \sim 10^{19}$ GeV, the Plank mass. Interestingly enough, this model give rise to observable effects in the behavior of entangled neutral meson systems, as shown also in [12], that can be experimentally tested.

KLOE has already published competitive results on these issues [13], based on a statistics of $\sim$400 pb^{-1}, and is now on the way of updating them using the full data sample.

The CPT symmetry breaking at KLOE could be also observed using kaon interferometry with both kaons decaying in $\pi^+\pi^-$ pairs. The refined analysis with additional data have brought the results on decoherence and CPT-violating parameters presented in Ref. [9]. A general theoretical possibility for CPT violation is based on spontaneous breaking of Lorentz symmetry, as developed by Kostelecky [14–16] in the *Standard-Model Extension* (SME). KLOE has already performed a preliminary analysis [9] using only half of the total available statistics (1 fb^{-1}). In this analysis the forward-backward asymmetry along Z direction as a function of sidereal time only was used. With this approach only the vector part of the Δa_μ four-vector can be observed. The preliminary results on $\Delta a_{X,Y,Z}$ are:

$$\begin{aligned}\Delta a_X &= (-6.3\pm6.0)\times10^{-18}\ \text{GeV},\\ \Delta a_Y &= (2.8\pm5.9)\times10^{-18}\ \text{GeV},\\ \Delta a_Z &= (2.4\pm9.7)\times10^{-18}\ \text{GeV}.\end{aligned}$$

A new analysis of the $K_LK_S \to \pi^+\pi^-\pi^+\pi^-$ final state is in progress to improve the SME parameters, through the measurement of the decay amplitude as a function of the delay between kaon decays, ordered according to the quadrant in the celestial coordinate frame [17]. The reconstruction of events in the region $\Delta t \approx 0$ is crucial for the precision measurement of the interference term.

5. KLOE–2 project

In 2011 the KLOE–2 detector, a successor of KLOE, has started data taking campaign in order to extend KLOE physics program [18].

The test of fundamental symmetries and Quantum Mechanics coherence of the neutral kaon system, and the search for phenomena beyond the Standard Model are the main motivations for KLOE–2. Thanks to the luminosity upgrade of DAΦNE, as well as ongoing installation of new detectors, KLOE–2 will be able to improve the accuracy of the measurement of the K_S meson properties and to study the time evolution of the entangled pairs of neutral kaons with an unprecedented precision. The significant improvement of the sensitivity of the tests of the discrete symmetries in the decays of K, η and η' mesons beyond the presently achieved limits is expected. Good accuracy of reconstruction of K_L decays in large fiducial volume was a base for the physics program of KLOE. At KLOE–2 an increased interest will be focused on the physics close to the interaction point (IP) as rare K_S decays, $K_S - K_L$ interference, multi-lepton events, as well as η, η' and $K^\pm$ decays. Detailed KLOE–2 physics program can be found in Ref. [18].

In 2011 the new interaction region of electron and positron beams consists on the Crabbed Waist compensation of the beam-beam interaction with large Piwinski angle and small beam sizes was successfully commissioned [19]. Upgraded DAΦNE allows to increase delivered luminosity (with unchanged beam current) by a factor of three with respect to the performance reached before the upgrade.

Apart of the collider the detector itself was also upgraded. Two pairs of small angle tagging devices [20] to detect low (Low Energy Tagger - LET) [21] and high (High Energy Tagger - HET) energy e^+e^- originated from $e^+e^- \to e^+e^-X$ reactions was installed. This tagger system allows to perform the studies of $\gamma\gamma$ physics. In the next step a light-material Inner Tracker detector based on the Cylindrical GEM technology will be installed in the region between the beam pipe and the drift chamber to improve charged vertex reconstruction and to increase the acceptance for low transverse momentum tracks [22, 23]. In addition, crystal calorimeters (CCALT) will cover the low polar θ angle region to increase acceptance for very forward electrons and photons down to 8° [24]. A new tile calorimeter (QCALT) will be used for the detection of photons coming from K_L decays in the drift chamber [25].

Figure 5 presents the comparison between results obtained with present spatial resolution and after installation of Inner Tracker for decoherence studies. An improved sensitivity on decoherence parameters of about one order of magnitude is expected with higher collected statistics and the use of Inner Tracker.

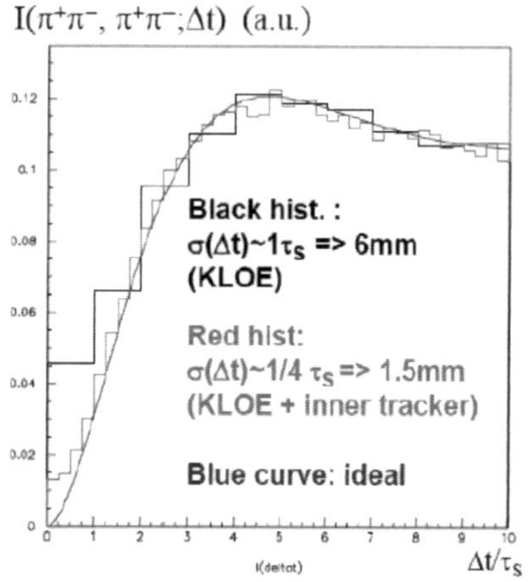

Figure 5. Comparison of the $I(\pi^+\pi^-, \pi^+\pi^-, \Delta t)$ distribution obtained with KLOE resolution and after Inner Tracker insertion based on the Monte Carlo simulation.

6. Summary

The KLOE experiment obtained already several significant and precise results in kaon physics [5], due to the unique possibility of producing pure K_L, K_S and $K^\pm$ beams. The success of the DAΦNE upgrade was a motivation to start a new experiment, KLOE–2, which aims to complete and extend the KLOE physics program. LET and HET tagging systems have been already installed and the first phase of new data taking campaign has started. After installation of new set of detectors (namely Inner Tracker, QCALT and CCALT calorimeters) new phase of data taking will start in 2013 for precise measurements of rare decays of kaon, η and η', as well as CP and CPT tests and kaon interferometry studies. The accuracy in the measurements is expected in most cases to be improved by about one order of magnitude [18].

Acknowledgments

The author would like to express his appreciation to Giorgio Giardina and Giuseppe Mandaglio for invitation and hospitality. The author acknowledges the support of the Polish National Science Centre through the Grant No. 2011/01/D/ST2/00748 and the Foundation for Polish Science through the project HOMING PLUS BIS/2011-4/3.

References

[1] *Handbook on neutral kaon interferometry at a Phi-factory*, editor: A. Di Domenico, Frascati Physics Series **43** (2007).
[2] A. Einstein, B. Podolsky, N. Rosen, Physical Review **47**, 777 (1935).
[3] The ϕ-factory Study Group *Proposal for a ϕ-factory*, LNF-90/031 (R) (1991).
[4] P. Franzini and M. Moulson, Ann. Rev. Nucl. Part. Sci. **56**, 207 (2006).

[5] F. Bossi,E. De Lucia,J. Lee-Franzini, S. Miscetti, M. Palutan and KLOE Collaboration, Rivista del Nuovo Cimento Vol.31, N.10 (2008).
[6] M. Adinolfi *et al.* (KLOE), Nucl. Instr. & Meth. **488**, 51 (2002).
[7] M. Adinolfi *et al.*, Nucl. Instr. & Meth. A **482**, 364 (2002).
[8] M. Silarski, PhD Thesis, Cracow (2012).
[9] A. Di Domenico *et al.* [KLOE Collaboration], J. Phys. Conf. Ser. **171**, 012008 (2009).
[10] J. Bernabeu, J. Ellis, N.E. Mavromatos, D.V. Nanopoulos and J. Papavassiliou, arXiv:hep-ph/0607322, in Ref. [1].
[11] J. Ellis, J.S. Hagelin, D.V. Nanopoulos, M. Srednicki, Nucl. Phys. B **241**, 381 (1984).
[12] P. Huet, M. Peskin, Nucl. Phys. B **434**, 3 (1995).
[13] F. Ambrosino et al., (KLOE Collaboration) Phys. Lett. B **642**, 315 (2006).
[14] V.A. Kostelecký, Phys. Rev. Lett. **80**, 1818 (1998).
[15] V.A. Kostelecký, Phys. Rev. **D61**, 016002 (1999).
[16] V.A. Kostelecký, Phys. Rev. **D64**, 076001 (2001).
[17] A. De Santis, PoS (HQL 2012) 018 (2012).
[18] G. Amelino-Camelia *et al.*, Eur. Phys. J. C **68**, 619-681 (2010).
[19] M Zobov *et al.*, Phys. Rev. Lett. **104**, 174801-174806 (2010).
[20] F. Archilli *et al.*, Nucl. Instr. & Meth. A **617**, 266 (2010).
[21] D. Babusci *et al.*, Nucl. Instr. & Meth. A **617**, 81 (2010).
[22] KLOE-2 Collaboration, F. Archilli *et al.*, LNF-10/3(P) INFN-LNF, arXiv:1002.2572
[23] A. Balla *et al.*, Nucl. Instr. & Meth. A **604**, 23 (2009).
[24] M. Cordelli *et al.*, Nucl. Instr. & Meth. A **617**, 109 (2010).
[25] M. Cordelli et al., Nucl. Instr. & Meth. **A 617**, 105 (2010).

Strong and Electromagnetic Interactions in High Energy Collisions IOP Publishing
Journal of Physics: Conference Series **424** (2013) 012006 doi:10.1088/1742-6596/424/1/012006

U boson search in the $e^+ + e^- \to \mu^+\mu^-\gamma$ process

Francesca Curciarello

on behalf of KLOE/KLOE2 Collaboration ‡

Dipartimento di Fisica e di Scienze della Terra, Università di Messina, via F.S. D'Alcontres, 98166 Messina, Italy

Istituto Nazionale di Fisica Nucleare, Sezione Catania, 95123, Catania, Italy

E-mail: `fcurciarello@unime.it`

Abstract. Following recent puzzling astrophysical results and recent theoretical studies, a search for a relatively low mass (1 GeV) new vector gauge boson (called the U boson) was performed by means of the KLOE detector. Investigations were carried out using the Initial State Radiation and searching for the signal from the U boson in the $e^+e^- \to \mu^+\mu^-\gamma$ process. The KLOE experiment, at the ϕ−factory DAΦNE (INFN-LNF), is the first to have exploited ISR to precisely determine cross sections of $e^+e^- \to \pi^+\pi^-\gamma(\gamma)$ and $e^+e^- \to \mu^+\mu^-\gamma(\gamma)$ processes below 1 GeV. Investigations were based on the data sample collected in 2002, which corresponds to integrated luminosity of 239.29 pb^{-1}. No evidence was found and a preliminary upper limit in the mass range between 600 and 1000 MeV was extracted.

‡ D. Babusci, D. Badoni, I. Balwierz-Pytko, G. Bencivenni, C. Bini, C. Bloise, F. Bossi, P. Branchini, A. Budano, L. Caldeira Balkeståhl, G. Capon, F. Ceradini, P. Ciambrone, F. Curciarello, E. Czerwiński, E. Dané, V. De Leo, E. De Lucia, G. De Robertis, A. De Santis, P. De Simone, A. Di Domenico, C. Di Donato, R. Di Salvo, D. Domenici, O. Erriquez, G. Fanizzi, A. Fantini, G. Felici, S. Fiore, P. Franzini, P. Gauzzi, G. Giardina, S. Giovannella, F. Gonnella, E. Graziani, F. Happacher, L. Heijkenskjöld, B. Höistad, L. Iafolla, E. Iarocci, M. Jacewicz, T. Johansson, K. Kacprzak, W. Kluge, A. Kupsc, J. Lee-Franzini, F. Loddo, P. Lukin, G. Mandaglio, M. Martemianov, M. Martini, M. Mascolo, R. Messi, S. Miscetti, G. Morello, D. Moricciani, P. Moskal, S. Muller, F. Nguyen, A. Passeri, V. Patera, I. Prado Longhi, A. Ranieri, C. F. Redmer, P. Santangelo, I. Sarra, M. Schioppa, B. Sciascia, M. Silarski, C. Taccini, L. Tortora, G. Venanzoni, R. Versaci, W. Wiślicki, M. Wolke, J. Zdebik

1. Introduction

The idea of the existence of a hidden gauge sector weakly coupled with the Standard Model (SM) through some mixing mechanism dates back to the early 80's. As is well known in fact, all our knowledge about the nature and composition of the Universe (see Fig. 1) represents only the 4-5% of what really exists, the remaining 95% is an unknown form of matter and energy. Particularly astrophysical data reveal that the 25% of

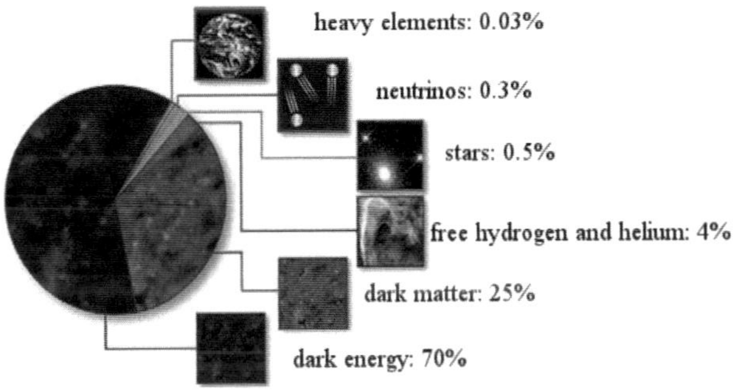

Figure 1. Universe composition.

Universe is made up of "dark matter" that, differently from the matter of atoms, does not emit or absorb light and has only been detected by its gravity. The "dark energy", instead, represents the 70% of Universe composition, this energy that permeates the universe acts as a sort of an anti-gravity and is responsible for the acceleration of the Universe expansion. For this reason, in the last decades, one of the main goal of particle physicists and astrophysicists has been to shed light on the existence of this unknown form of matter and energy. The experiments are mainly oriented towards the discovery of new particles at high energy scales, however, a complementary interesting approach is the search for new light particles at relatively low energy scales. Such particles may have remained undiscovered because of their weak coupling to the SM.

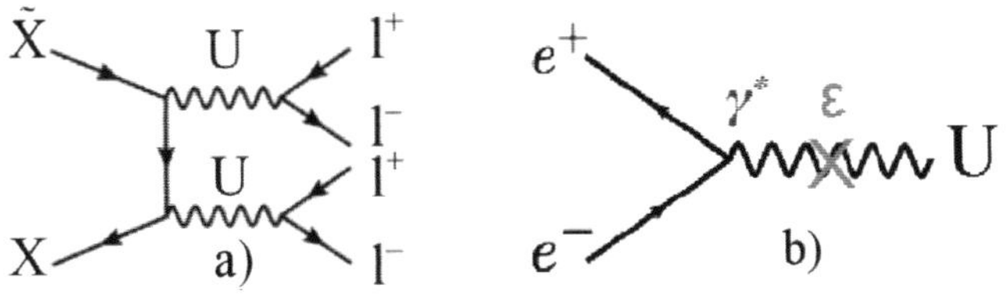

Figure 2. a) U boson production by dark matter particles annihilation; b) Coupling of U boson to SM particles.

Moreover, in the last two years striking astrophysical observations, that cannot be interpreted by standard astrophysics and particle physics, have attracted particular attention of the Scientific Community on the study of a hidden low–energy dark sector. The most important are: the $e^+ - e^-$ excess in the cosmic ray flux and the absence of a similar effect in proton/anti-proton observations by PAMELA [1]; the 511 keV gamma ray signal form galactic center by INTEGRAL satellite [2]; the DAMA/LIBRA annual modulation signal [3]. All these observations could be explained if one assumes that a dark force gauge boson, mediator of an unknown dark force with mass less than two proton mass, $M_U < 2\,m_p$, exists.

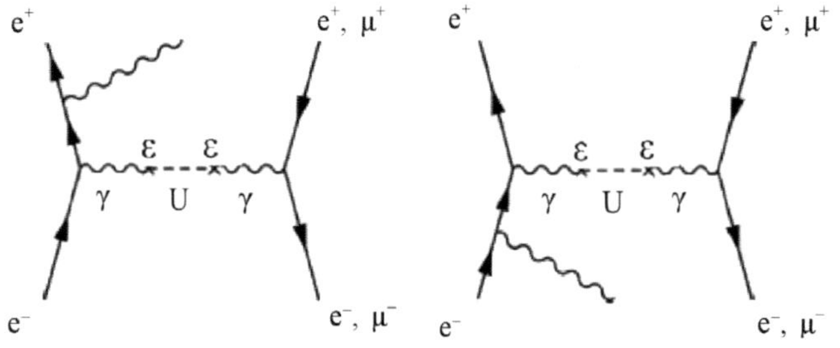

Figure 3. Examples of Feynman diagrams with dark photon exchange contributing to $e^+e^- \to e^+e^-(\mu^+\mu^-)\gamma$ process.

The U boson also called dark photon is predicted by several Standard Model Extensions (SME) [4–9]. According to dark force models this dark force gauge boson would be produced during dark matter annihilation processes and then decay into light particles, as leptons, ($\tilde{X}+X \to U+U$, $U \to l^+l^-$ $l = e, \mu$, see Fig. 2a)), assuming a mass less than 1 GeV [10]. Such dark photon is associated to an abelian gauge symmetry that can communicate with the ordinary SM through a kinetic mixing term that is given by [4, 5, 10, 11] :

$$L_{mix} = -\frac{\varepsilon^2}{2} F_{ij}^{e.m.} F_{dark}^{ij} \tag{1}$$

where $\varepsilon^2 = \alpha'/\alpha$ is the kinetic mixing parameter ($\alpha = 1/137$, α' is the U boson coupling constant, see Fig. 2b)), $F_{ij}^{e.m.}$ is the electromagnetic tensor, F_{dark}^{ij} is the dark matter hypercharge gauge boson tensor.

Luckily, dark force models make a number of predictions that can be tested by particle physics experiments. Particularly, high luminosity $e^+ - e^-$ collider experiments at GeV scale can be a direct probe of Dark Forces. At flavor factories a particular clean channel is the production of the U boson plus a photon with the consequent decay of the U boson in a lepton pair: $e^+e^- \to U\gamma \to l^+l^-\gamma$, where $l = e, \mu$ (see Fig 3) [10]. The expected U boson signal should have the shape of a Breit-Wigner peak in the invariant mass distribution of the lepton pair, as shown in Fig.4). For this reason, about 240 pb^{-1} of KLOE data taken in 2002 were used to analyse the $e^+e^- \to \mu^+\mu^-\gamma$ channel with the aim to search for the light vector gauge boson.

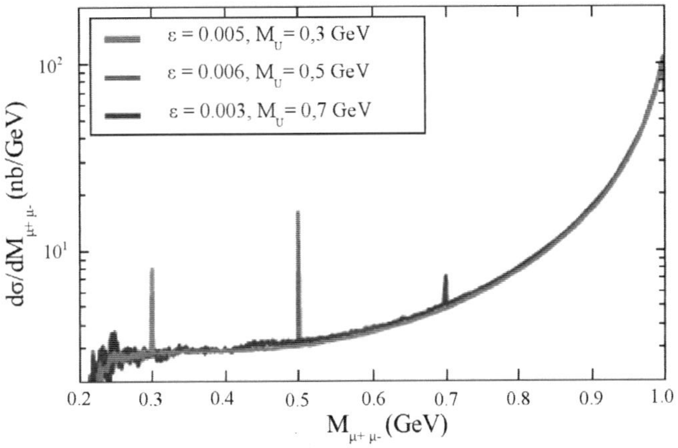

Figure 4. Simulation of invariant mass distribution of the muon pairs with U boson peak for different values of the kinetic mixing parameter and U boson mass, at a Φ–factory with $\sqrt{s} = 1.02\,GeV$ [10].

2. KLOE Experimental Set-Up

DAΦNE is a $e^+ - e^-$ collider working at the energy $\sqrt{s} = m_\phi = 1.0195$ GeV and it is located at the INFN-LNF of Frascati. The DAΦNE Accelerator Complex consists of a linear accelerator, a damping ring, nearly 180 m of transfer lines, two storage rings that intersect at two points, a beam test area (BTF) and three synchrotron light lines (see Fig.5).

Figure 5. DAΦNE accelerator complex.

The KLOE detector is made up of a large cylindrical drift chamber (DC, see Fig.6 left side), surrounded by a lead scintillating fiber electromagnetic calorimeter (EMC, see Fig.6 right side). A superconducting coil around the EMC provides a 0.52 T magnetic field. The EMC provides measurement of photon energies, impact point and an accurate measurement of the arrival time of particles. The DC is well suited for tracking of the particles and charged vertices reconstruction. The calorimeter is divided into a

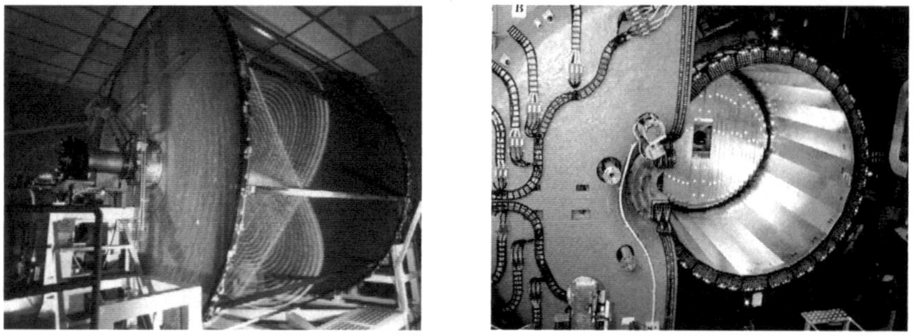

Figure 6. left side : KLOE drift chamber; right side: KLOE calorimeter.

barrel and two end–caps and covers 98% of the complete solid angle. The modules are read out at both ends by 4880 photo–multipliers. Energy and time resolutions are $\sigma_E/E = 5.7\%/\sqrt{E(GeV)}$ and $\sigma_t = 57\,ps\ /\sqrt{E(GeV)} \oplus 50\,ps$, respectively. The all-stereo drift chamber, $4\,m$ in diameter and $3.3\,m$ long, is made of carbon fiber-epoxy composite and operates with a light gas mixture (90% helium, 10% isobutane). The position resolutions are about $\sigma_{xy} = 150\,\mu m$ and $\sigma_z = 2\,mm$. The momentum resolution is $\sigma_{p_\perp}/p_\perp > 0.4\%$ for large angle tracks. Vertices are reconstructed with a spatial resolution of about $3\,mm$.

3. Data Analysis

The analysis described in the following is based on the well grounded analysis developed for the KLOE pion form factor measurement [12]. Since for this particular search a direct correspondence between ε^2 and U-boson mass was needed, differently from the analysis of reference [12] the $\mu^+\mu^-\gamma$ absolute cross section was derived as function of the invariant mass.

The search is based on the data sample taken in the year 2002 corresponding to the integrated luminosity of $239.29\,pb^{-1}$. The selection requires:

- two charged tracks with $50° < \theta_\mu < 130°$ (wide cones in Fig. 7)
- one photon within a cone of $\theta_\gamma < 15°$ ($\theta_\gamma > 165°$) around the beamline (narrow cones in Fig. 7).

The photon is not detected, its direction is reconstructed from event kinematics: $\vec{p}_\gamma \simeq \vec{p}_{miss} \equiv -\vec{p}_{\mu\mu} = -(\vec{p}_{\mu^+} + \vec{p}_{\mu^-})$. This separation of tracks and photon selection regions in the analysis, greatly reduces the contamination from the resonant processes $e^+e^- \to \phi \to \pi^+\pi^-\pi^0$, where charged pions are misidentified as muons and the π^0 mimics the missing momentum of the photon(s), and from the final state radiation processes: $e^+e^- \to \pi^+\pi^-\gamma_{FSR}$ and $e^+e^- \to \mu^+\mu^-\gamma_{FSR}$. Since ISR-photons are mostly collinear with the beam line, the applied angular cut does not decrease high statistics of signal events.

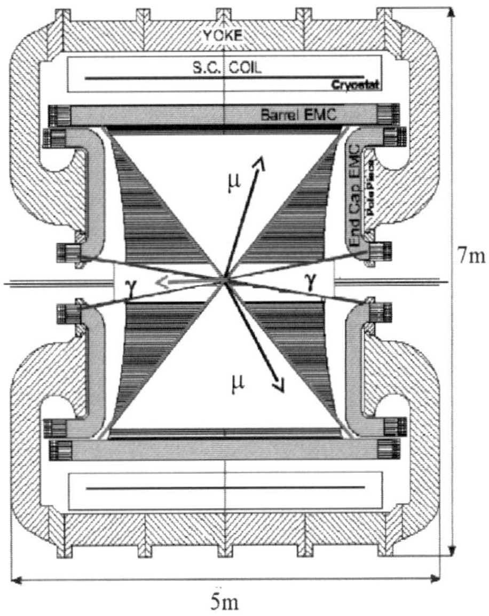

Figure 7. Schematic cross-view of KLOE Detector with the selections regions for the missing momentum angle (narrow cones) and for muon tracks (wide cones).

Background contributions coming from channels:

1) $e^+e^- \to \pi^+\pi^-\gamma(\gamma)$

2) $e^+e^- \to \Phi \to \pi^+\pi^-\pi^0$

3) $e^+e^- \to \mathrm{e}^+\mathrm{e}^-\gamma(\gamma)$

were separated applying kinematical cuts in the $M_{\mathrm{T}rk} - M^2_{\pi\pi}$ plane, (for the definition of M_{Trk} variable see later), as it is possible to see in Fig 8a). A particle identification estimator (PID), based on a pseudo-likelihood function using the time-of-flight and calorimeter information, was used to suppress radiative Bhabha events [12–14]. Finally, as shown in Fig. 8b), pions and muons were separated by a cut on the variable $M_{\mathrm{T}rk}$: particularly muons are selected with $80 < M_{\mathrm{T}rk} < 115$ MeV while pions are selected with $M_{\mathrm{T}rk} > 130$ MeV. The M_{Trk} variable is computed from energy and momentum conservation, assuming the presence of an unobserved photon and that the tracks belong to particles of the same mass:

$$\left(\sqrt{s} - \sqrt{|\mathbf{p}^+|^2 + M^2_{\mathrm{Trk}}} - \sqrt{|\mathbf{p}^-|^2 + M^2_{\mathrm{Trk}}}\right)^2 - \left(\mathbf{p}^+ + \mathbf{p}^-\right)^2 = 0 \qquad (2)$$

where $\mathbf{p}^\pm$ is the measured momentum of the positive (negative) particle, and only one of the four solutions has physical meaning. As it is shown in Fig. 8b) the gap between the two selections has been chosen to reduce the mutual contamination of the two samples.

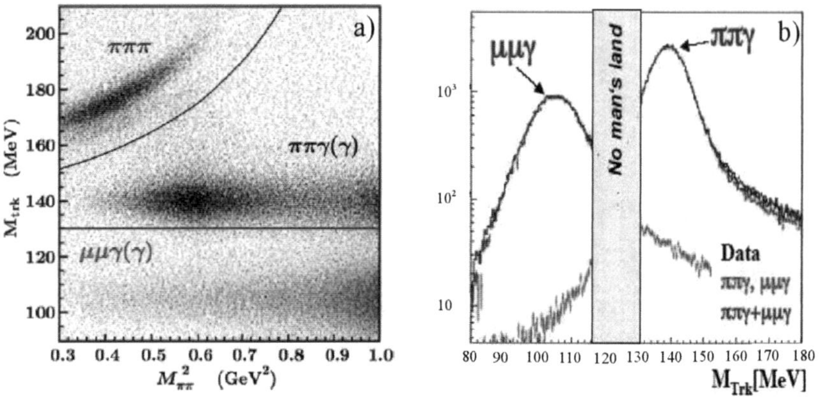

Figure 8. a) MC simulation of signal and background distributions in the $M_{Trk} - M^2_{\pi\pi}$ plane; b) $\mu^+\mu^-\gamma$ and $\pi^+\pi^-\gamma$ separation by cut on M_{Trk} distribution; data are represented in black, the MC simulation of $\pi^+\pi^-\gamma$ and $\mu^+\mu^-\gamma$ distributions is in red, in blue is shown their sum.

3.1. Background contributions

The main background channels surviving the $\mu^+\mu^-\gamma$ selection cuts are due to the 1), 2), and 3) channels. Their contributions are obtained by fitting in slices of $M^2_{\mu\mu}$, the M_{Trk}

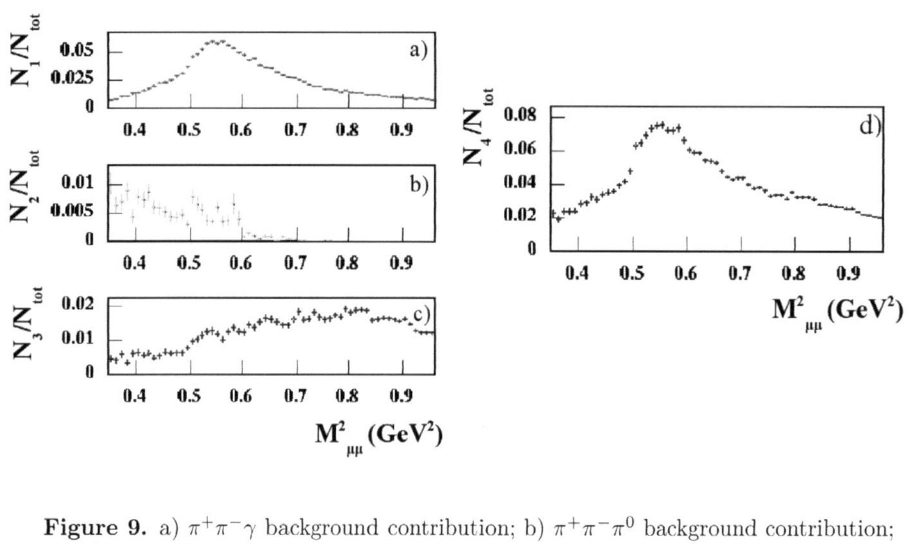

Figure 9. a) $\pi^+\pi^-\gamma$ background contribution; b) $\pi^+\pi^-\pi^0$ background contribution; c) $e^+e^-\gamma$ background contribution; d) sum of the three relative background contributions a), b) and c). N_1, N_2, N_3 are the background event numbers related to the above mentioned 1), 2), 3) channels, respectively; N_4 represents their sum.

distribution for data as a superposition of signal and background distributions. For each background channel, the fit results in normalization parameters, called weights, $w_{ch}(j)$, produced for each slice of $M^2_{\mu\mu}$. The $M_{\mathrm{T}rk}$ distribution for $\mu^+\mu^-\gamma$, $\pi^+\pi^-\gamma$, $\pi^+\pi^-\pi^0$ were simulated, whereas the $e^+e^-\gamma$ contribution was estimated based on data. The reason is that the $e^+e^-\gamma$ residual background is much larger, at percent level, and therefore must be carefully evaluated. Once the normalization parameters $w_{ch}(j)$ are obtained in each slice of $M^2_{\mu\mu}$, they are used to find the background contribution for the $\mu^+\mu^-\gamma$ signal region defined by $80 < M_{Trk} < 115MeV$, as shown in Fig. 9a), b) and c). The fraction of background events (N_1, N_2, N_3), was obtained in each bin of $M^2_{\mu\mu}$ with respect to the number of data events N_{tot} found in the same bin. The obtained fractional background contribution (see Fig. 9d)) was then used in the background subtraction procedure. Since in the ρ region the $\pi^+\pi^-\gamma$ contribution is about one order of magnitude greater than the $\mu^+\mu^-\gamma$ one, it is crucial to to keep under control the $\pi^+\pi^-\gamma$ $M_{\mathrm{T}rk}$ tail in the region below 125 MeV. For this reason, a tuning of the $M_{\mathrm{T}rk}$ tail using a control sample of $70\,pb^{-1}$ of $\Phi \to \pi^+\pi^-\pi^0$ was also applied.

3.2. Cut on the σ_{MTRK}

In order to improve the π/μ separation a continuous function cut on the quality of fitted tracks parametrized as σ_{MTRK}, was applied to data. Since σ_{MTRK} value depends on the quality of the fitted tracks, bad reconstructed tracks are expected to contribute to large values of σ_{MTRK}.

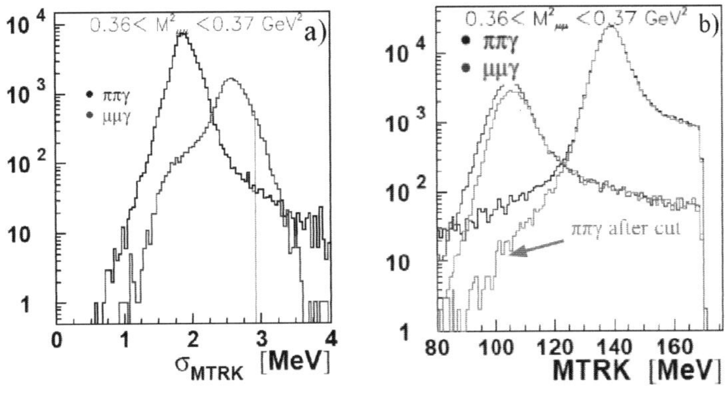

Figure 10. a) σ_{MTRK} for $\pi^+\pi^-\gamma$ (black) and $\mu^+\mu^-\gamma$ (blue) is represented for an exemplary $M^2_{\mu\mu}$ slice. In red a possible cut value is shown; b) black and blue histograms represent the $\pi^+\pi^-\gamma$ and $\mu^+\mu^-\gamma$ $M_{\mathrm{T}rk}$ distributions without σ_{MTRK} cut, red histograms are obtained after applying σ_{MTRK} cut.

In Fig. 10a) are reported the $\pi^+\pi^-\gamma$ (black) and $\mu^+\mu^-\gamma$ (blue) σ_{MTRK} distributions for one slice of $M^2_{\mu\mu}$. Figure 10b) shows the effects of σ_{MTRK} cut (red) on $M_{\mathrm{T}rk}$ distribution for one slice of $M^2_{\mu\mu}$. As expected , there is a significant reduction (up to a factor 2) of the $\pi^+\pi^-\gamma$ contamination in the $\mu^+\mu^-\gamma$ $M_{\mathrm{T}rk}$ region, with a consequent improvement in π/μ separation. The cut was optimized in order to keep the signal

($\mu^+\mu^-\gamma$) at a level of about 70% .

3.3. $\mu^+\mu^-\gamma$ Absolute Cross Section

Once the Data/MC corrections have been applied, the $\mu^+\mu^-\gamma$ cross section was extracted by subtracting the residual background to the observed spectra and dividing it by efficiencies and the integrated luminosity. The $\mu^+\mu^-\gamma$ absolute cross section, derived

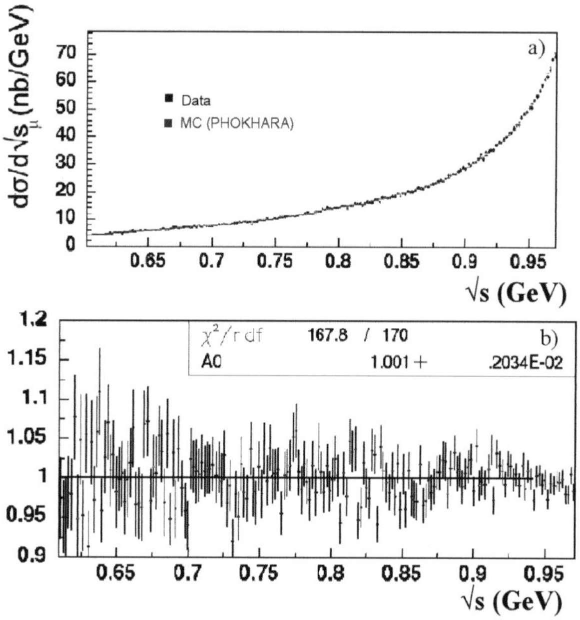

Figure 11. a) Comparison of data (black points) and MC (blue points) of $\mu^+\mu^-\gamma$ absolute cross section; b) ratio of Data to MC PHOKHARA prediction.

by the analysis procedure described above, was then compared with the NLO QED MC prediction from PHOKHARA Event Generator [15]. The comparison is shown in Fig. 11. As it is possible to see an excellent agreement between Data and PHOKHARA MC prediction was achieved.

4. Upper Limit Extraction on ε^2

To extract the upper limit (U.L.) on ε^2 the TLimit Root Class [16], based on the confidence level signal (CLS) technique [17] was used. To compute the limit the observed spectrum, that is the raw spectrum after offline background filter efficiency corrections (which are at percent level) and background subtraction, was used as "data" input of TLimit procedure. As "background" input was used the MC PHOKHARA spectrum properly normalized to the raw spectrum. In Fig. 12 the exclusion plot on number of signal events at 90% of confidence level (C.L.) in the 600 – 1000 MeV energy range

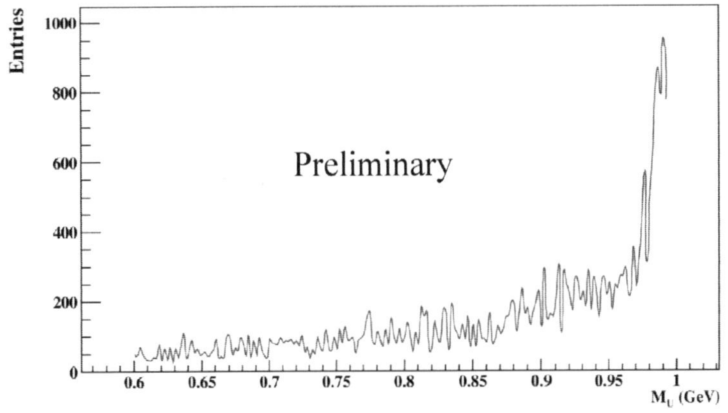

Figure 12. Exclusion plot on number of signal events at 90% of C.L..

is shown. A systematic error of $\sim$ 2% on background was also applied to the TLimit procedure. The U.L. on the kinetic mixing parameter was extracted using the following formula [18]:

$$\varepsilon^2 = \frac{\alpha'}{\alpha} = \frac{N_{CLS}/(\epsilon_{eff} \cdot L)}{H \cdot I} \tag{3}$$

where N_{CLS} is the number of entries of signal hypothesis excluded as fluctuations at the 90% C.L.; ϵ_{eff} represents the acceptance and efficiency corrections; L is the integrated luminosity ($L = 239, 29\,pb^{-1}$); H is the radiator function given by:

$$H = \frac{d\sigma_{\mu^+\mu^-\gamma}/d\sqrt{s_\mu}}{\sigma(e^+e^- \to \mu^+\mu^-, s_\mu)} \tag{4}$$

where $d\sigma_{\mu^+\mu^-\gamma}/d\sqrt{s_\mu}$ is the partial cross section of $e^+e^- \to \mu^+\mu^-\gamma(\gamma)$, s_μ is the invariant mass of muons, $\sigma(e^+e^- \to \mu^+\mu^-, s_\mu)$ is the total cross section of $e^+e^- \to \mu^+\mu^-$ process; the term I is given by the following integral:

$$I = \int_i \sigma_U^{\mu\mu}\, ds_i \tag{5}$$

where $\sigma_U^{\mu\mu} = \sigma(e^+e^- \to U \to \mu^+\mu^-, s)$ is the total cross section of U boson production decaying in the $\mu^+\mu^-$ channel, $s = M_U^2$, i is the mass bin number. The result of the application of this formula is given in Fig.13. In this figure the exclusion plot on the kinetic mixing parameter ε^2 determined in this analysis is compared to the other existing limits. The blue area shows the present measurement derived using the $e^+e^- \to \mu^+\mu^-\gamma$ channel, the U.L. is between $2.6 \cdot 10^{-6}$ and $3.5 \cdot 10^{-7}$, it is clearly visible the reduction of the sensitivity due to ρ meson at about $0.77\,GeV$. The red area represents the Mami results [19], the dark green area the Apex measurement [20] and finally, the blue shaded area is the KLOE U.L. in the 0$-$460 MeV region calculated using the Dalitz Φ decay [21]. The black line represents the ε^2 values consistent with a U boson contribution to the muon magnetic moment anomaly a_μ.

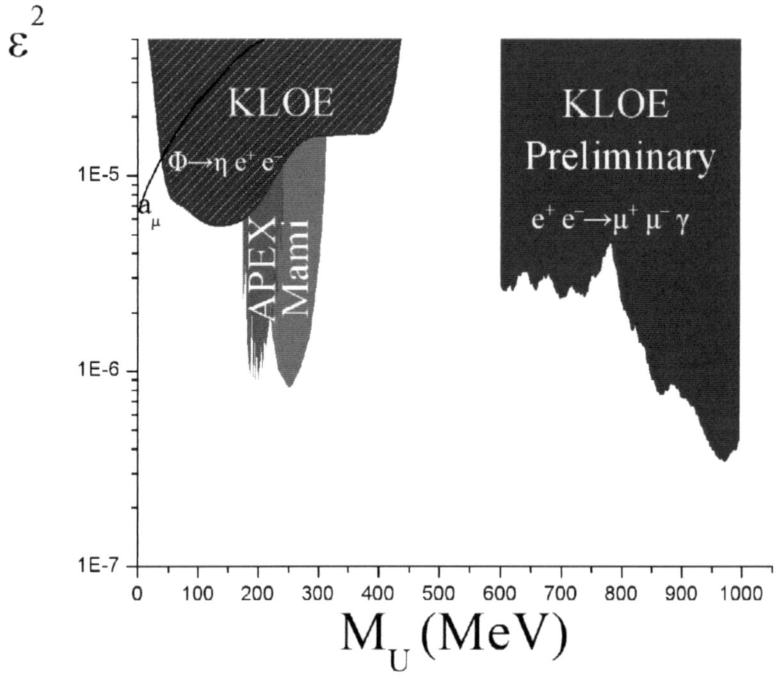

Figure 13. Exclusion Plot on ε^2. The preliminary result of this analysis (blue area) is compared to the existing limits in the range $0 - 1000\,MeV$.

5. Conclusions

KLOE data collected in the year 2002 at $\sqrt{s} = 1\,GeV$ (corresponding to the luminosity of $239.29\,pb^{-1}$) were used to search for light vector boson in the $e^+e^- \to \mu^+\mu^-\gamma$ channel. No U boson evidence was found and an U.L. has been extracted on coupling factor ε^2 in the energy range between 600 and $1000\,MeV$. The presented preliminary results exclude a possible effect of U boson existence on the muon magnetic moment anomaly a_μ in the energy range between 600 and 1000 MeV. In future an analysis of full KLOE data sample ($2.5\,fb^{-1}$) is planned in order to extend the search to the lower invariant masses of muon pair and in order to increase a sensitivity by a factor of about 3.

References

[1] Adriani O *et al.* 2009 *Nature* **458** 607
[2] Jean P *et al.* 2003 *Astronomy Astrophysics* **407** L55
[3] Barnabei R *et al.* 2008 *Eur. Phys. J* C **56** 333
[4] Boehm C and Fayet P 2004 *Nucl. Phys.* B **683** 259
[5] Pospelov M *et al.* 2008 *Phys. Lett.* B **662** 53
[6] Pospelov M *et al.* 2009 *Phys. Lett.* B **671** 391
[7] Arkani-Hamed N *et al.* 2009 *Phys. Rev.* D **79** 015014.
[8] Cholis I *et al.* 2009 *Phys. Rev.* D **80** 123518.
[9] Mambrini Y and Cosmol J 2010 *Astropart. Phys.* **1009** 022

[10] Barzè L *et al* 2011 *Eur. Phys. J* C **71** 1680
[11] Holdom B 1986 *Phys. Lett.* B **166** 196
[12] Babusci D *et al.* arXiv:1212.4524 [KLOE/KLOE2 Collaboration]
[13] Ambrosino F *et al.* 2009 *Phys. Lett.* B **670** [KLOE Collaboration] 285
[14] Ambrosino F *et al.* 2011 *Phys. Lett.* B **700** [KLOE Collaboration] 102
[15] H. Czyż *et al.* 2005 *Eur. Phys. J* C **39** 411
[16] http://root.cern.ch/root/html/TLimit.html
[17] Feldman G C and Cousins, R D 1998 *Phys. Rev.* D **57** 3873
[18] Venanzoni G, KLOE2 Internal Note K2ID-12,
`http://www.lnf.infn.it/kloe2/index.php?lv1=documents&lv2=list&lv3=private&lv4=k2id`
[19] Merkel M *et al.* 2011 *Phys. Rev. Lett.* **106** 251802
[20] Abrahamyan S *et al.* 2011 *Phys. Rev. Lett.* **107** 191804
[21] Babusci D *et al.* arXiv:1210.3927 [KLOE/KLOE2 Collaboration]

Journal of Physics: Conference Series **424** (2013) 012007 doi:10.1088/1742-6596/424/1/012007

Meson Spectroscopy at JLab@12 GeV

Andrea Celentano
Physics Department, Genoa University, IT
E-mail: andrea.celentano@ge.infn.it

Abstract. Meson, being the simplest hadronic bound system, is the ideal "laboratory" to study the interaction between quarks, to understand the role of the gluons inside hadrons and to investigate the origin of color confinement. To perform such studies it is important to measure the meson spectrum, with precise determination of resonance masses and properties, looking for rare $q\bar{q}$ states and for unconventional mesons with exotic quantum numbers (i.e. mesons with quantum numbers that are not compatible with a $q\bar{q}$ structure).
With the imminent advent of the 12 GeV upgrade of Jefferson Lab a new generation of meson spectroscopy experiments will start: "Meson-Ex" in Hall B and "GLUEX" in Hall D. Both will use photo-production to explore the spectrum of mesons in the light-quark sector, in the energy range of few GeVs.

1. Introduction

The phenomenology of hadrons and in particular the study of their spectrum led more than forty years ago to the development of the quark model, where baryons and mesons are described as bound systems of three quarks and of a quark-antiquark pair, respectively. While this picture still holds and has been proven to reproduce many features of the hadron spectrum, now we know that the hadron mass cannot be explained only in terms of the quark masses, but it is mainly due to the dynamics of the gluons that bind them. Measuring the spectrum of hadrons, studying their properties and their inner structure is therefore crucial to achieve a deep knowledge of the strong force.

Mesons, being made by a quark and an anti-quark, are the simplest quark bound system and therefore the ideal "laboratory" to study the interaction between quarks, understand the role of gluons, and investigate the origin of confinement. The Constituent Quark Model predicts the existence of multiplets of mesons with similar properties and masses, that are classified according to their total angular momentum J, the parity P, and charge conjugation C. While most of the lowest mass states have been clearly identified and studied [1], several open issues related to the mass hierarchy and decays of excited states remain and still await for a experimental investigation. In addition phenomenological models [2, 3, 4, 5, 6, 7] and lattice QCD calculations [8, 9] suggest that states beyond the simple $q\bar{q}$ configuration, as hybrids (qqg), tetraquarks ($qq\bar{q}\bar{q}$), and glueballs, should also exist. If it would be the case, a much richer spectrum than that predicted by the quark model should be expected and, in particular, new meson multiplets corresponding to these unconventional configurations should be observed.

Hybrid mesons are of particular interest as they are the cleanest experimental signature for the presence of gluons in the dynamical mass generation process. A precise determination of their spectrum and properties can provide a unique opportunity to study the role of the glue and to understand the phenomenon of confinement.

An unambiguous identification of these states can in general be rather difficult, since they can mix with ordinary mesons having the same quantum numbers (J^{PC}). However, the additional degrees of freedom present in these states can also lead to *exotic* quantum numbers that are forbidden in $q\bar{q}$ systems, and therefore provide a unique signature of their unusual structure.

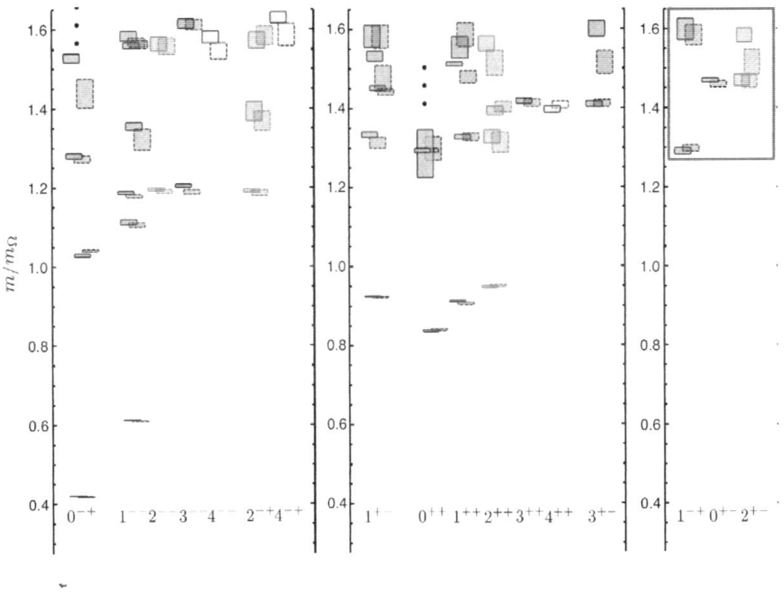

Figure 1. Recent lattice results from JLab lattice collaboration [9] for the low-lying isovector mesons. Masses are given in units of the Ω^- baryon. States within the red box have *exotic* quantum numbers.

2. Contemporary Meson Spectroscopy Experiments

As per today the experimental determination of unconventional mesons is rather confused, since there are not clear, unambiguous, and well shared evidences of their existence. Most of the data on exotic mesons in the light quark sector have been collected in the last thirty years with hadronic probes, either in diffractive production with pion beams or anti-proton annihilation on protons and neutrons [10, 11, 12].

The most solid signal appears to be the $\pi_1(1600)$ ($J^{PC} = 1^{-+}$), that has been observed in several decay modes, namely $\eta'\pi$, $b_1\pi$, $f_1\pi$, $\rho\pi$, and in several experiments. The signal reported by the E852 Collaboration in the $\pi^- p \to \pi^-\pi^-\pi^+ p$ channel [13] is considered the most robust evidence for this state. However, a later analysis of the same channel performed with more statistics and more precise analysis tools [14], as well as negative observations from VES [15] and CLAS [16], raised strong doubts on the existence of this state. The situation is made even more complicated by the recent positive result reported by the COMPASS collaboration [17].

The situation for the other two possible states that have been reported, the $\pi_1(1400)$ and the $\pi_1(2105)$, is even more controversial (see [12] and [11] for a complete review).

To solve this complicate "puzzle" it is necessary to work on three main points. First, an exotic signal to be convincing has to be found in more than one decay mode, since the background that can affect the signal extraction is different, giving different sensitivity to the signal. Second, exotics need to be searched for in experiments using different probes, since the dominant production mechanisms are still unknown. Finally, precise and robust analysis tools have to be used, since the final results are strictly dependent on their reliability.

A new generation of meson spectroscopy experiments is just started and more will come in the near future, with the ultimate goal of solving the unconventional mesons "puzzle". Such experiments will collect large datasets of high quality data exploiting different production mechanisms: COMPASS at Cern [18] will use the diffractive production by a 200 GeV pion beam on both proton and nuclear targets, BES-III in Beijing [19] will use e^-e^+ annihilation, PANDA at GSI [20] will use proton-antiproton annihilation, and finally GlueX [21] and MesonEx [22] at JLab will use photoproduction on a proton target.

2.1. Amplitude Analysis

The goal of a meson spectroscopy program is to identify resonances measuring their decay products. A resonance is formally described as a pole in the production amplitude with defined angular momentum and isospin.

In practice, resonances are numerous, often broad and overlapping each other: only for a narrow and well-isolated state, like the ρ, the resonant structure can be identified by looking at the invariant mass spectrum of the decay products. The identification of a precise state requires the extraction of the corresponding waves from the measured distributions.

This task is performed via the "Partial Wave Analysis" technique (PWA): the cross section of the process is parametrized as a coherent sum of different amplitudes, with defined quantum numbers. The cross section is then fitted to the data to extract amplitudes. Fits can be performed as a function of the invariant mass of the measured decay products and other relevant kinematic variables to derive information on the dependence of the amplitudes on these variables. The PWA is a crucial component for a meson spectroscopy program since the final results are strictly dependent on the reliability of the PWA tools and of the theoretical assumption used in constructing the waves. From the technical side, the improvement in computer science has removed the most part of the limitations that were common in the past. This will allow the analysis of large data sets with adequate number of waves in a reasonable time. From the theoretical side, partial wave analysis can be improved by simplifying the fits introducing known constraints on the amplitudes, and including basic and fundamental properties, such as analiticity and unitarity.

3. Photoproduction of hybrid and exotic mesons

Phenomenological models, like the flux-tube model, indicate that the photon may be more effective in producing exotics hybrids in diffractive production on a proton target than, for example, the pion. The reason for this lies in the fact that the photon spin is one and it can fluctuate into a $q\bar{q}$ pair with spins aligned. When a $q\bar{q}$ pair with $\vec{S} = 1$ is excited into a hybrid, the production of exotic quantum numbers is expected to be favored. Phenomenological studies also indicate that exotic mesons can be produced with photon probes with cross sections comparable to ordinary mesons [23, 24]. Finally, the linear polarization of the photons is expected to be a very powerful tool for high precision amplitude analysis, since it can provide information on the production mechanism and acts as a filter for background.

4. Meson Spectroscopy at Jefferson Laboratory

4.1. The Jefferson Laboratory National Accelerator Facility

The Thomas Jefferson National Accelerator Facility (TJNAF or JLab) houses the Continuous Electron Beam Accelerator Facility (CEBAF), a high-current, high-duty electron beam machine with three experimental Halls. A schematic of the accelerator site is shown in figure 2. The CEBAF is composed by two linear accelerators, each adding to the electrons up to 0.4 GeV per pass, and two sets of recirculating arcs. The electron beam circulates up to five times in the accelerator, to reach the maximum energy of about 6 GeV. Three independent beam, with different energies and currents, can be sent to the the three experimental halls. The polarization of the electron beam can be as high as 85% and it is pseudo-randomly flipped every about 30 ms to reduce systematic effects.

JLab is currently undergoing an upgrade of the accelerator and of the existing experimental Halls, while a new experimental Hall (Hall D) is being constructed, as shown in figure 3. The accelerator portion of the upgrade will be constructed on the framework of the existing CEBAF accelerator. Five new superconducting radio-frequency accelerating elements will be added to each of the Linacs, and the existing RF cavities will be increased in gradient to achieve a 1.1 GeV/linac accelerating power, and a new recirculating arch will be added to provide an extra pass through the North Linac. Such new configuration will bring the beam up to 11 GeV for Halls A, B and C and up to 12 GeV for Hall D.

Figure 2. Aerial view of the JLab accelerator site.

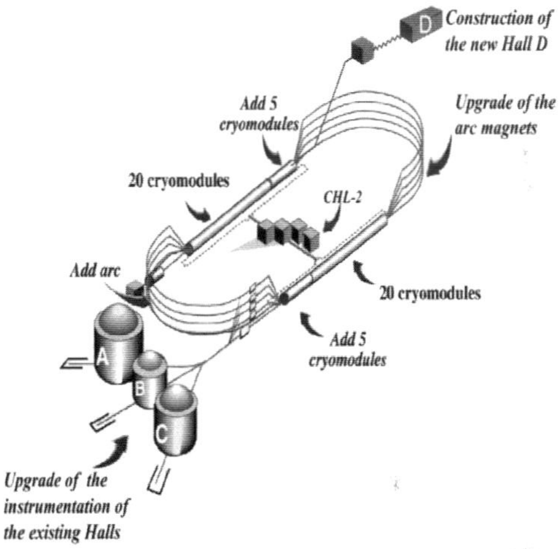

Figure 3. The JLab accelerator site after the 12 GeV upgrade.

4.2. Meson Spectroscopy Experiments at JLab

Meson Spectroscopy is one of the main topics that will be studied at JLab in the 12 GeV era, with two dedicated experiments: GlueX in Hall D and MesonEx in Hall B. The scientific goal of both experiments is to perform a comprehensive measurement of the meson spectrum in the light quark sector, with precise determination of resonance masses and properties, looking for rare $q\bar{q}$ states and for unconventional mesons with exotic quantum numbers. Both experiments will use photon beams with high intensity and linear polarization on a proton target. Final state particles will be measured with large acceptance detectors: GlueX and CLAS12.

5. The GlueX experiment in Hall D

The GlueX experiment will be held at JLab in the newly-constructed Hall D, with a new dedicated detector. It will run a meson spectroscopy program using high-energy, linearly polarized photons on a proton target.

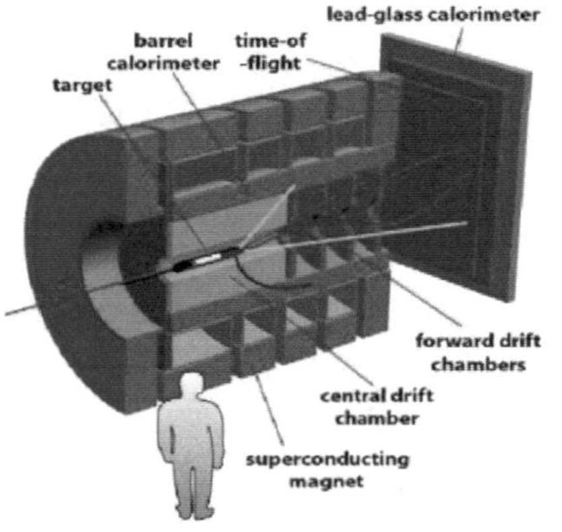

Figure 4. The GlueX detector that will be installed in Hall D at JLab.

Figure 5. The CLAS12 detector that will be installed in Hall B at JLab.

5.1. The GlueX detector

The GlueX detector is based on a solenoid, that is ideal for a fixed-target photoproduction experiment (figure 6). The solenoidal magnetic field traps low energy electromagnetic backgrounds (e^+e^- pairs) generated in the target inside a cone around the beam. It also allows for effective instrumentation of calorimeters to achieve very high acceptance for photons. The superconducting solenoid produces a 2.25 T field. A tagged, high energy, linearly polarized ($\simeq 40\%$) Bresshtrahlung photon beam impings on a 30 cm long liquid-hydrogen target surrounded by a start counter to trigger the DAQ. Moving from the inner to the outer layers of the detector, there is a cylindrical tracking chamber, the CDC, and a cylindrical electromagnetic calorimeter, the BCAL. Downstream of the CDC are four sets of circular planar drift chambers, FDC, followed by a time-of-flight wall, TOF. This is followed by a circular planar electromagnetic calorimeter, the FCAL. Some space has been reserved between the downstream end of the magnet and the TOF for a possible particle identification (PID) system. This design provides for nearly 4π acceptance for both charged particles and photons. While the acceptance is not completely uniform in all variables, there are no holes in the kinematic variables of interest.

5.2. The Gluex photon beam

The GlueX experiment requires a photon beam with energies in the range 8-9 GeV, high linear polarization and high intensity (up to $10^8\gamma/s$), that has to be obtained from the primary electron beam delivered by the CEBAF accelerator. The method used to produce such a beam is coherent Bremsstrahlung in a oriented crystal. 12 GeV electrons from the accelerator pass through a thin diamond radiator ($20\mu m$), generating an intense beam of high-energy photons with a continuous energy spectrum dominated by a single peak: figure 7 shows the typical Bremsstrahlung spectrum calculated for a diamond crystal radiator of 20 μm and a 1 μA electron beam of 12 GeV. A significant fraction of the total power in the beam is concentrated inside the peak. Using electrons with energy E_0 the peak position can be varied, from 0 to 90% E_0 simply rotating the crystal.

The photon spectrum inside the peak has a large degree of linear polarization: figure 8 shows the linear polarization of the photons in the peak as a function of the energy of the electron beam. The entire beam polarization appears in the coherent peak, while the continuous background only dilutes it. To get maximum polarization it is crucial to separate the two contributions and take

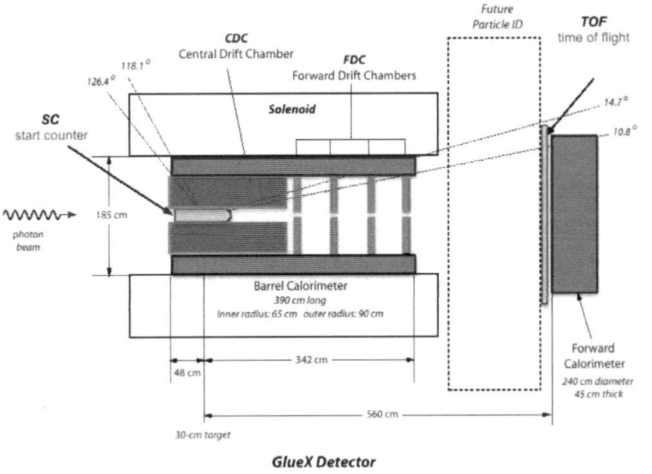

Figure 6. Different subsystems of the GlueX detector. In orange, the charged particle tracking subsystem. In blue, the calorimetry subsystem. In green, the particle identification subsystem. See text for the description of the different subdetectors.

only photons inside the peak. This can be done using a suitable collimator, exploiting the fact that the two components have angular distributions with different characteristics. Collimating away all photons beyond some angle θ, the incoherent spectrum is attenuated uniformly at all energies, while the coherent spectrum under the relevant peak is left unchanged from the maximum energy down to a certain cut-off.

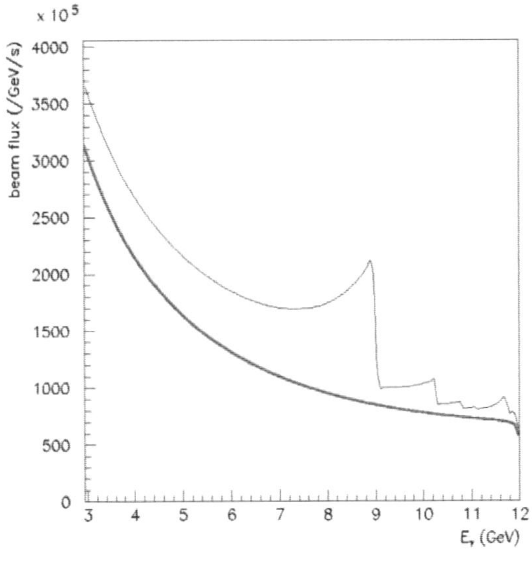

Figure 7. Bremsstrahlung spectrum for a diamond crystal radiator of $20\mu m$ and a 1 μA electron beam of 12 GeV. The coherent peak is clearly visible. The red line represents the contribution of the incoherent Bremsstrahlung. Figure is taken from [25].

The precise energy of each produced photon is determined measuring the momentum of the corresponding recoil electron in a dedicate, Elbek-type tagging spectrometer, schematically shown in figure 9.

The primary 12 GeV electrons from the CEBAF accelerator hit the diamond radiator, where a small fraction of them emits Bremsstrahlung photons. Then the electrons enter in a quadrupole

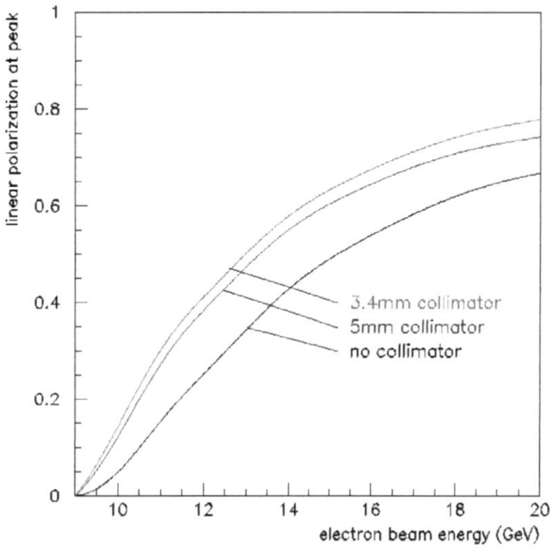

Figure 8. The linear polarization of the photons under the coherent peak of the Bremsstrahlung spectrum reported in figure 7. The black curve refers to the case without collimator, while red and blue curves refer to the case with 2 different collimators. Figure is taken from [25].

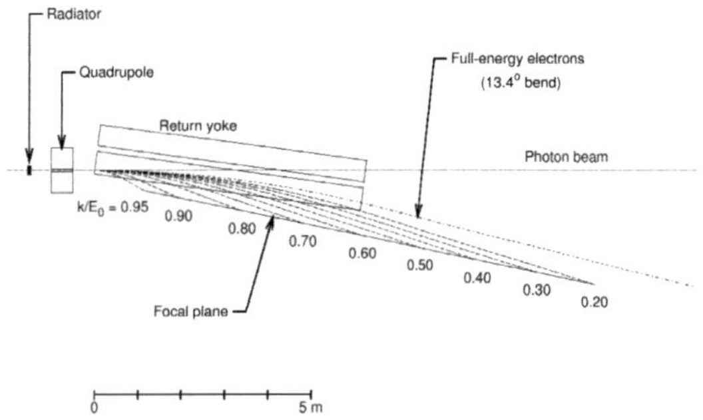

Figure 9. Schematic view of the Gluex photon tagging spectrometer, showing the path of the primary beam and of the electrons that emitted a Bremsstrahlung photon in the radiator, as a function of the emitted photon momentum k over the primary beam energy E_0. Figure is taken from [25].

and in the dipole field region, to be bended by the tagger magnet. Electrons that did not go under Bremsstrahlung are deflected by 13.4° going straight toward the beam dump. The others are deflected more, depending on the energy they lost in the radiator, entering in the active detector region, immediately after the focal plane in figure 9.

The spectrometer is made by two detectors: a set of 190 fixed scintillator counters spanning the electron momentum range from 3.0 to 11.7 GeV and a movable "microscope" of 500 scintillating fibers. With the fixed array it is possible to measure the full tagged photon spectrum, with energy resolution of 0.25 % and reduced rate. The movable "miscroscope" is used whenever a better resolution or a highest rate is required.

5.3. Expected performances and results

The reaction $\gamma p \to 3\pi\, n$ involving three charged pions in the final state has been used as a benchmark to test the potential of the GlueX detector and tools developed for the analysis. The reason why this reaction has been chosen is that it has been studied by past experiments, like E852 [13] and VES [15], with controversial results about the presence of the exotic $\pi_1(1680)$ state. A further investigation of this final state is therefore highly desirable.

The test procedure was the following: pseudo Monte-Carlo data have been generated using information from E852 as input, including the exotic $\pi_1(1600)$, simulating the statistic corresponding to 3.5 hours of beamtime in nominal operating conditions. Then this set of data

Journal of Physics: Conference Series **424** (2013) 012007 doi:10.1088/1742-6596/424/1/012007

was passed trough the Monte-Carlo code of the detector to simulate its response. Finally, an amplitude analysis based on a isobar model was performed, dividing the 3π invariant mass range in independent, 20 MeV width, bins.

The intensity for this reaction, having a photon beam along the z axis with fractional linear polarization f and including n different resonances X_β is written as follows:

$$I = \sum_{p=\pm 1} \frac{1+pf}{4} |\sum_{\beta=1}^{n} V_{p,\beta}(< X_\beta|R > +pe^{2i\alpha} < X_\beta|L >)|^2 \quad (1)$$

where $|R>$ and $|L>$ are the photon helicity basis states and α is the angle between the production plane of the resonance X and the plane of polarization of the photon beam. $V_{p,\beta}$ is the production amplitude for the resonance X_β, that is replaced by a free parameter in the maximum likehood fit involved in the partial wave analysis. Finally, $p = +1$ represents photon polarization along x and $p = -1$ represents photon polarization along y.

Table 1 reports the resonances included in the analysis.

Table 1. Resonances included in the 3π final state analysis.

Resonance	Decay channel	Relative angular momentum L
$a_1(1230)$	$\rho\ \pi$	S
$a_2(1320)$	$\rho\ \pi$	D
$\pi_2(1670)$	$f_2\ \pi$	S
$\pi_2(1670)$	$\rho\ \pi$	P
$\pi_1(1600)$	$\rho\ \pi$	P

Results are reported in figure 10. The signal from the $\pi_1(1600)$ is clearly visible, even if it corresponds only to 1.5% of the overall intensity. This demonstrates that the GlueX detector can be effectively used to perform a comprehensive partial wave analysis and find exotic signals.

6. The MesonEx experiment in Hall B

MesonEx (JLab Exp-11-005) is an experimental program that will study meson spectroscopy trough quasi-real photoproduction. It will take place in the Hall B of Jefferson Laboratory, using the CLAS12 detector [27] and a new Forward Tagger facility (FT). The goal of this experiment is to perform a comprehensive study of the meson spectrum in the light quark sector, in the energy range of few GeVs, with precise determination of resonance masses and properties.

6.1. Quasi-real photoproduction

During the 6 GeV operations, photoproduction experiments were run in Hall-B making use of the CLAS detector [26] for the produced hadrons and a Bremsstrhalung tagger facility to produce and tag real photons in the energy range between 1 GeV to 5.5 GeV. This setup can not be operated at 11 GeV because of the limited bending power of the existing magnet.

In the MesonEx experiment, quasi-real photons on an hydrogen target will be used. Photons are produced by the 11 GeV electron beam from the CEBAF accelerator scattering at very low angle (2.5°-4.5°) from the target, with Q^2 values of about $10^{-1}\,GeV^2$ or lower.

In the unpolarized electron scattering process (one-photon exchange approximation), the virtual photon *transverse* polarization is:

$$\varepsilon = [1 + 2\frac{Q^2+\nu^2}{Q^2}\tan^2(\theta_{e'}/2)]^{-1} \quad (2)$$

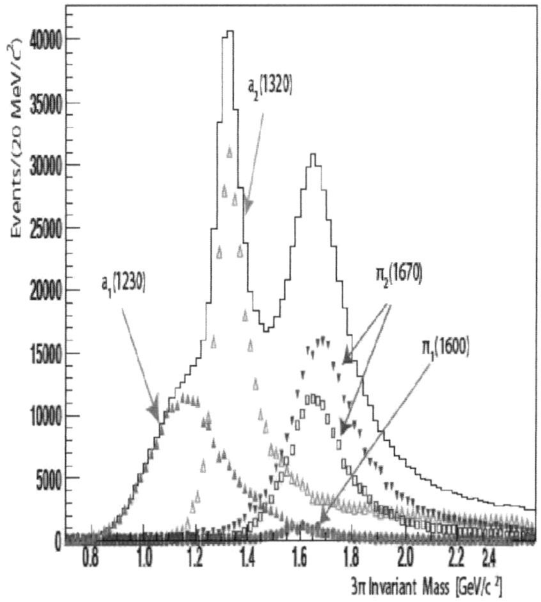

Figure 10. Result of the GlueX partial wave analysis of the 3π final state. The black curve reports the overall intensity for the reaction, while the colored points report the intensities of the 5 different waves for each 3π invariant mass bin.

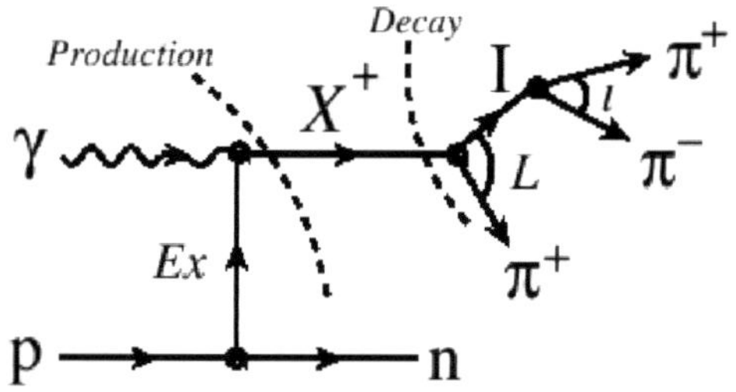

Figure 11. Isobar model for the reaction $\gamma p \to 3\pi n$. X is the produced resonance, that then decays into isobar I and a "spectator" pion. I in turn decays into 2 pions.

where ν is the photon energy and $\theta_{e'}$ the electron scattering angle. The *longitudinal* polarization is given by $\varepsilon_L = \frac{Q^2}{\nu^2}\varepsilon$ [28]: at very low values of Q^2 the virtual photon beam becomes, for all practical purposes, almost a real photon beam, since $\varepsilon_L \simeq 0$.

Using the above relations the photon polarization will be determined event by event, simply measuring the electron three-momentum. The polarization plane coincides with the electron scattering plane and the degree of polarization mainly depends on the energy: the associated systematic uncertainty is only affected by the electron detection resolution.

In the MesonEx experiment the CLAS12 detector will be used to detect hadrons in the final state. However, its geometrical acceptance prevents to detect electrons at low angles: it is therefore necessary to add a new component, the Forward Tagger, to measure electrons scattered below 5° and thus to reconstruct kinematic variables of the quasi-real photons.

6.2. The CLAS12 detector

As part of the 12-GeV upgrade of the CEBAF accelerator complex at Jefferson Lab, the existing CLAS detector in Hall B will be upgraded to CLAS12. This detector is designed to perform experiments using high energy electron beam inpinging on polarized and unpolarized targets at luminosities up to $L = 10^{35}\, cm^{-2}\, s^{-1}$. CLAS12 consists of two parts, a forward detector (FD) and a central detector (CD), respectively shown in figure 12 and 13.

The Forward Detector will detect and identify charged and neutral particles scattered between 5° and 40°, in the full momentum range. Particles will be detected in six identical magnetic spectrometers based on six coils, superconducting toroidal magnet. Each spectrometer will be equipped with a Forward Vertex Tracker (FVT), high threshold Cherenkov Counter (HTCC),

Journal of Physics: Conference Series **424** (2013) 012007 doi:10.1088/1742-6596/424/1/012007

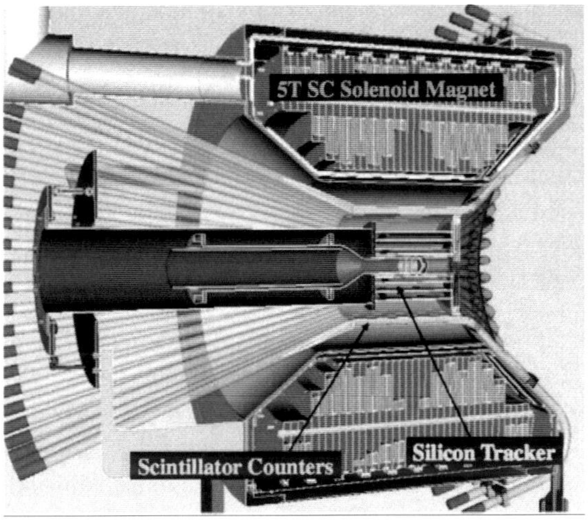

Figure 12. The CLAS12 Central Detector. See text for a complete description of subsystems.

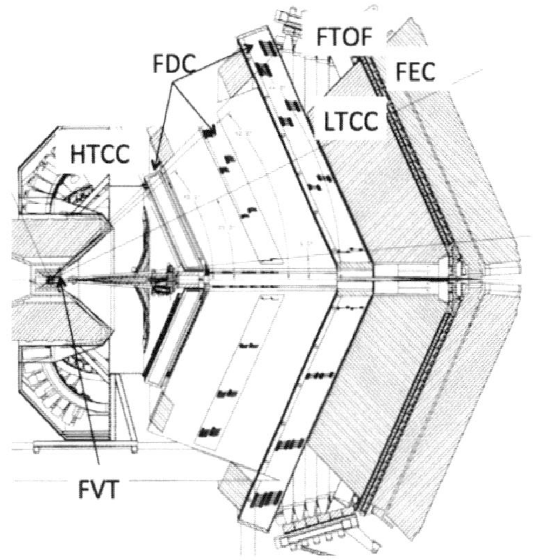

Figure 13. The CLAS12 Forward Detector. See text for a complete description of subsystems.

a set of Drift Chambers (FDC), low threshold Cherenkov Counter (LTCC), time-of-flight scintillator counters (FTOF), and electromagnetic calorimeters (FEC).

The Central Detector is meant to measure charged hadrons and neutrons in the angular range from 35° to 125° and momenta lower than 1.5 GeV/c. Charged particles will be detected by two barrel shaped detectors, Si-tracker (BST) and time-of-flight counters (CTOF). Both detectors are positioned inside a solenoidal magnet and cover the full azimuthal angular range. Neutrons will be detected by a dedicated detector (CND) positioned between CTOF and the cryostat of the solenoid.

6.3. The Forward Tagger detector

The angular range of interest for scattered electron detection in the MesonEx experiment is between 2.5° and 4.5°, corresponding to a Q^2 range between 0.01 GeV^2 and 0.3 GeV^2. This angular range is outside the CLAS12 acceptance, therefore a new dedicated detector, the Forward Tagger, has to be added to the experimental setup.

To reconstruct the quasi-real photon variables it is necessary to measure the scattered electron three momentum. The most relevant quantities are its energy $E_{e'}$, its polar angle $\varphi_{e'}$, and its azimuthal angle $\theta_{e'}$. From them it is possible to reconstruct the quasi-real photon energy ν, the Q^2, the polarization ε, and the polarization plane, that coincides with the electron scattering plane.

The Forward Tagger will be made of a *calorimeter* (FT-Cal), to identify the electron, measure its energy, and provide a fast trigger signal, a *tracker* (FT-Trck), to measure the scattering angles, and a *scintillation counter* (FT-Hodo) to provide e/γ separation. The Forward Tagger will be placed between the high threshold Cherenkov Counter (HTCC) and the torus support, at about 190 cm downstream of the target (nominal) position.

The FT-Cal is the core of the Forward Tagger. It has to fulfill strong requirements in terms of radiation hardness, light yield, radiation length, recovery time, energy and time resolution. Due to the limited space available and its position inside CLAS12 it has to be compact and operable in a high magnetic field environment.

The choice that has been done in the MesonEx experiment is to construct a calorimeter based on $PbWO_4$ scintillating crystals. In the recent years this material has been extensively studied and shown to be very resistant to radiation damage. $PbWO_4$ crystals has been used in large scale detectors, such as CMS-ECal[29], ALICE-PHOS [30], PANDA-EMC [31] and CLAS-IC [32]. $PbWO_4$ has a very fast scintillation decay time (6.5 ns), a very small radiation lenth (0.9 cm), and a small Moliere radius (2.1 cm). Even if the light yield is only 0.3% of NaI(Tl) the designed energy resolution of the overall calorimeter is of $2\%/\sqrt{E(GeV)} \oplus 1\%$.

Due to the expected high rage from electromagnetic background, the calorimeter will be highly segmented in the transverse direction to maintain each channel at a sustainable readout-rate. The size of each crystal will be $1.5 \times 1.5\,cm^2$, comparable with the characteristic transverse size of the electromagnetic shower to contain the signal induced by incident electrons to few crystals.

The crystal size is also well matched to the photodetectors. For each crystal the light will be read-out by an Avalanche Photo Diode (APD Hamamatsu S8664-1010), with active area of $10 \times 10\,mm^2$. This design has been shown to meet the required criteria in terms of radiation hardness and operability in a magnetic field.

6.4. Expected performances and results

The reaction $\gamma p \to 3\pi n$, involving three charged pions in the final state, has also be used as a benchamrk reaction to study the performances of the MesonEx experimental setup and the analysis tools to perform the Partial Wave Analysis. As already described for GlueX, it is also a reaction with large physical interest due to the contradictory results about the $\pi_1(1600)$ reported by different past experiments in this channel.

This reaction can be easily accessed in CLAS12 by detecting the three charged pions in the forward part of the CLAS12 Detector. The exclusivity of the reaction is ensured by using the Forward Tagger to determine the 3-momentum of the initial state photon and then applying the missing mass technique to select events with a missing neutron.

The experimental acceptance for this channel has been evaluated generating events according to the Isobar model with a flat 3π invariant mass and exponential t channel distribution, $\frac{d\sigma}{dt} \propto e^{-5|t|}$. The response of CLAS12 and FT was then analized performing a full simulation of the detector.

The most important kinematic variables for this reaction, that the production amplitude may depend from, are the invariant mass of the 3π system, the momentum transfer t and the polar angles of the isobar in the meson resonance decay, as measured in the Gottfried-Jackson (GJ) frame. Figure 14 shows the acceptance for these kinematics variables, for a magnetic field configuration corresponding to half of the nominal value. This configuration has been chosen due to the higher acceptance at low $M_{3\pi}$ (approximately a factor 5 at 1.4 GeV/c^2 respect to the nominal field configuration).

The suitability of the MesonEx experimental setup and analysis tools for Partial Wave Analysis has been tested as follows. First a set of pseudo Monte-Carlo data have been generated according to a specifically-designed model for the 3π production [22]. The model defines 8 possible final states, summarized in table 2, decaying to $\pi^+\pi^+\pi^-$, with an additional recoiling neutron in the final state. An exotic $\pi_1(1600)$ state is included, with a contribution of 2% to the overall intensity. One million of events have been generated, divided into two t bins at -0.2 and -0.5 GeV^2, each with the same number of events.

These events were then tracked through the detector via a MonteCarlo code, and only those with 3 reconstructed pions were accepted for the PWA. Finally, maximum likehood fits were performed independently in each 3π invariant mass bin to extract production amplitudes.

Figure 15 shows the comparison between the generated waves and the result of the fits for each t bin. The generated waves are reproduced very well for all channels, including the π_1 exotic. Even if its contribution is just 2% of the total intensity, a clear, statistically significant

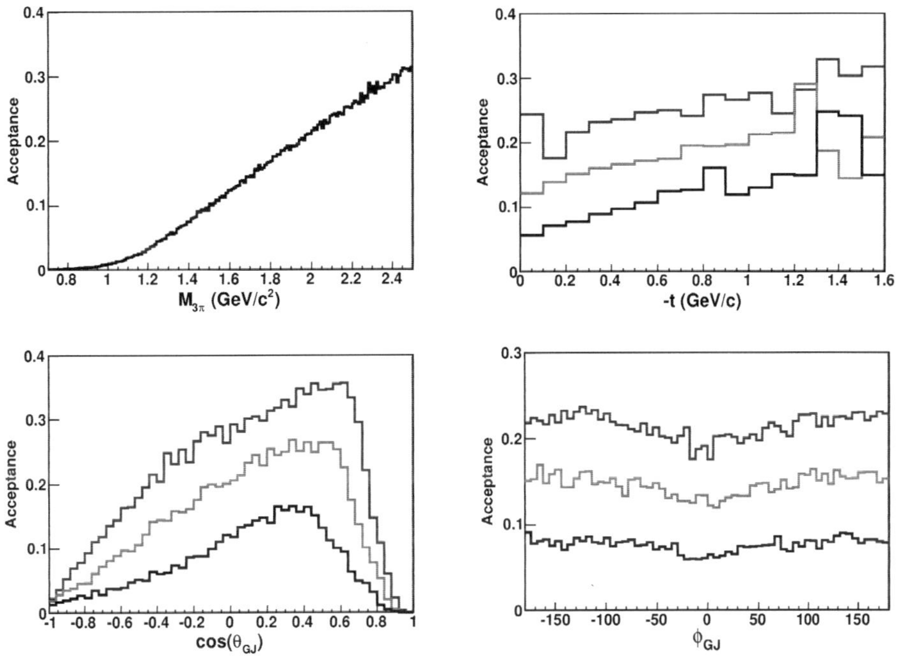

Figure 14. Acceptances for the $\pi^+\pi^+\pi^-n$ final state and magnetic field set to half of the nominal value. Top left the 3π invariant mass $M_{3\pi}$, right the momentum transfer t, bottom left, $\cos\theta_{GJ}$ and right, ϕ_{GJ}. For the latter 3 plots the black, red and blue lines correspond to $M_{3\pi} = 1.4$, 1.7 and 2.0 (GeV/c^2) respectively.

signal is reproduced.

This leads to the conclusion that the CLAS12-Forward Tagger system is intrinsically capable of meson spectroscopy measurements via Partial Wave Analysis.

Table 2. The meson states produced in the model of $\pi^+\pi^+\pi^-$ production, detailed in [22]. L is the relative orbital angular momentum between the isobar and the bachelor pion.

State	J^{PC}	L	Decay Mode
$a_1\,(1260)$	1^{++}	D	$\rho\pi$
$a_2\,(1320)$	2^{++}	D	$\rho\pi$
$\pi_2\,(1670)$	2^{-+}	P	$\rho\pi$
$\pi_2\,(1670)$	2^{-+}	F	$\rho\pi$
$\pi_2\,(1670)$	2^{-+}	S	$f_2\pi$
$\pi_2\,(1670)$	2^{-+}	D	$f_2\pi$
$\pi_1\,(1600)$	1^{-+}	P	$\rho\pi$

7. Conclusions

Meson Spectroscopy is a powerful tool to answer to fundamental questions in QCD, as the origin of color confinement and the role of gluons inside hadrons. Mesons are the simples quark bound

Journal of Physics: Conference Series **424** (2013) 012007 doi:10.1088/1742-6596/424/1/012007

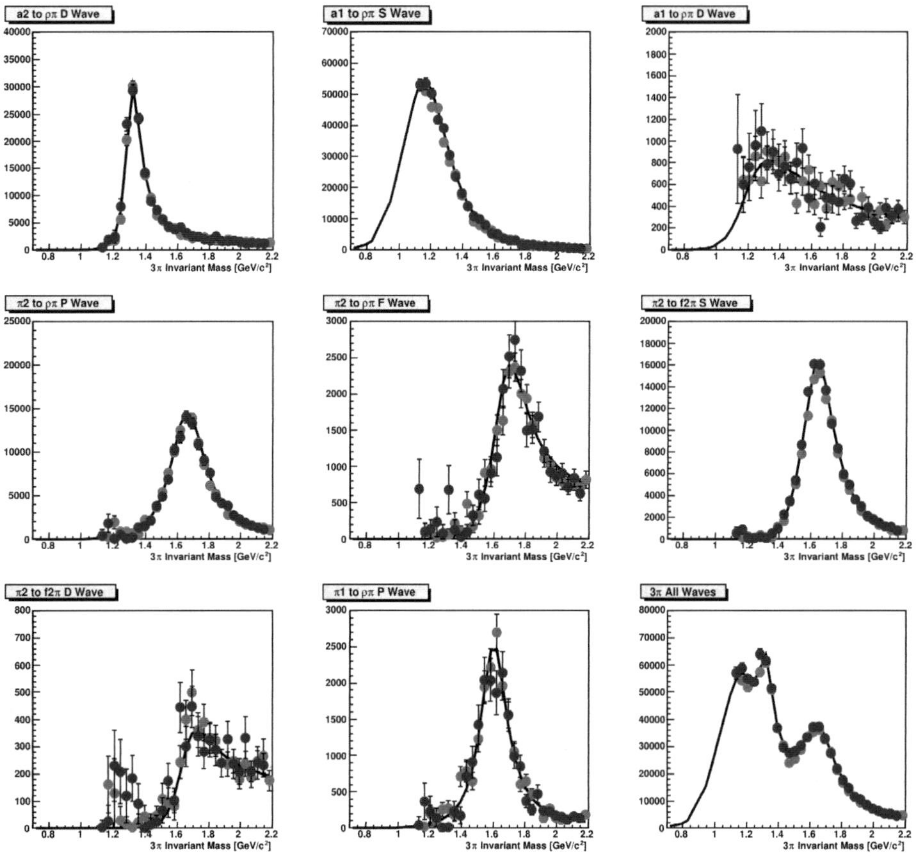

Figure 15. The intensities of the 8 different isobar channels in the 3π model, see table 2. The bottom right plot shows the total intensity. The black line shows the generated waves, while the blue and red points are the fit results for $t = 0.2$ and $0.5\ (GeV/c)^2$, respectively.

system, and therefore the ideal "laboratory" to study the strong force at the non-perturbative energy scale of few GeVs. In particular, unconventional mesons (i.e. states incompatible with a simple $q\bar{q}$ structure) would be the best experimental evidence of the active role of gluons in hadron dynamics.

A new generation of high precision meson spectroscopy experiments is starting in the near future with the ultimate goal of solving the unconventional mesons "puzzle", that has been delineated by the controversial results reported by past measurements. Jefferson Laboratory will play an active role in this program during the 12 GeV era, with two dedicated experiments: GlueX in Hall D and MesonEx in Hall B. They will be installed inside the experimental halls during 2014, to be ready for data taking in 2015.

The goal of these experiments is to use photo-production on a proton target to investigate the meson spectrum in the energy range of few GeVs, looking for rare $q\bar{q}$ and exotic states: Partial Wave Analysis techinque will be used to determine their mass and properties.

References

[1] K. Nakamura *et al J. Phys.* G **37**

[2] Jaffe R. L. and Johnson K. 1971 *Phys. Lett.* B **60** 201
[3] Horn D. and Mandula J. 1978 *Phys. Rev.* D **17** 898
[4] Barnes T., Close F. E., de Viron F. and Weyers J. 1983 *Nucl. Phys.* B **224** 241
[5] Isgur N., Kokoski R. and Paton J. E. 1985 *Phys. Rev. Lett.* **54** 869
[6] Isgur N. and Paton J. E. 1985 *Phys. Rev.* **D31** 2910
[7] Guo P., Szczepaniak A.P., Galatà G., Vassallo A. and Santopinto E. 2008 *Phys. Rev.* D **78** 056003
[8] Dudek J. J., Edwards R. G., Peardon M. J. *et al.* 2009 *Phys. Rev. Lett.* **103** 262001
[9] Dudek J. J., Edwards R. G., Peardon M. J. *et al.* 2010 *Phys. Rev.* D **82** 034508
[10] Godfrey S. and Napolitano J. 1999 *Rev. Mod. Phys.* **71** 1411-62
[11] Klempt E. [hep-ph/0404270].
[12] Meyer C. A. and Van Haarlem Y. 2010 *Phys. Rev.* C **C82** 025208
[13] Adams G. S. *et al.* (E852 Collaboration) 1998 *Phys. Rev. Lett.* **81** 5760
[14] Dzierba A. *et al.* 2006 *Phys. Rev.* D **73** 072001
[15] Khokhlov Y. *et al.* (VES Collaboration) 2000 *Nucl. Phys.* A **663** 596
[16] Nozar M. *et al.* (CLAS Collaboration) 2009 *Phys. Rev. Lett.* **102** 102002
[17] Alekseev A. *et al.* (COMPASS Collaboration) 2010 *Phys. Rev Lett.* **104** 241803
[18] Abbon P. *et al.* (COMPASS Collaboration) 2007 *Nucl. Instrum. Meth.* A **557** 455-518
[19] Ablikim M. *et al.* (BESIII Collaboration) 2010 *Nucl. Instrum. Meth.* A **614** 345
[20] Bettoni D. *et al.* (PANDA Collaboration) 2009 Physics performance report for PANDA: strong interaction studies with antiprotons *Preprint* arXiv:0903.3905v1 [hep-ex]
[21] The GlueX Collaboration Mapping the spectrum of light quark mesons and gluonic excitations with linearly polarized photons *Preprint*
[22] Battaglieri M. *et al.* 2011 Meson Spectroscopy with low Q^2 electron scattering in CLAS12 *Preprint*
[23] Afanasev A. and Page P. R. 1998 *Phys. Rev.* A **57** 6771
[24] Szczepaniak A. and Swat M. 2001 *Phys. Lett.* B **516**
[25] The GlueX Collaboration 2008 *The technical design of the Hall-D polarized photon beam at the Thomas Jefferson National Accelerator Facility*
[26] Mecking B. A. *et al.* (CLAS Collaboration) 2003 *Nucl. Instrum. Meth.* A **503**
[27] The CLAS collaboration *CLAS12 Technical design report*
[28] Dombey N. 1969 *Rev. Mod. Phys.* **41** 236
[29] CMS Collaboration 1997 *CMS Ecal Technical Design Report* (CERN-LHCC 97-33)
[30] Zhou D. C. *et al.* (ALICE Collaboration) 2007 *J. Phys.* G **34** 719
[31] PANDA Collaboration 2008 *Technical Design Report for PANDA Electromagnetic Calorimeter (EMC)*
[32] Bedlinskiy I. *et al.* *An Inner Calorimeter for CLAS/DVCS experiments*

Recent Results of the BGO-OD Experiment at ELSA Facility

Veronica De Leo[1,2] B. Bantes[3], D. Bayadilov[4], R. Beck[4], M. Becker[4], A. Bella[3], J. Bieling[3], S. Boese[4], A. Braghieri[5], K. Brinkmann[6], D. Burdeynyi[7], F. Curciarello[1,2], R. Di Salvo[8], H. Dutz[3], D. Elsner[3], A. Fantini[8,9], T. Frese[3], F. Frommberger[3], V. Ganenko[7], G. Gervino[10,18], F. Ghio[11,12], G. Giardina[1,2], B. Girolami[11,12], D. Glazier[13], S. Goertz[3], A. Gridnev[14], D. Hammann[3], J. Hannappel[3], W. Hillert[3], A. Ignatov[15], O. Jahn[3], R. Jahn[4], R. Joosten[4], T.C. Jude[3], F. Klein[3], K. Koop[4], B. Krusche[16], A. Lapik[15], P. Levi Sandri[17], I. Lopatin[14], G. Mandaglio[1,2], F. Messi[3], R. Messi[8,9], D. Moricciani[8], V. Nedorezov[15], D. Noviskiy[14], P. Pedroni[5], M. Romaniuk[1,2], T. Rostomyan[16], C. Schaerf[8,9], H. Schmieden[3], V. Sumachev[14], V. Tarakanov[14], V. Vegna[3], P. Vlasov[4], D. Walther[4], D. Watts[13], H.-G. Zaunick[4], T. Zimmermann[3]

[1] Università di Messina, Dipartimento di Fisica e di Scienze della Terra, via F.S.D'Alcontres, 98166 Messina, Italy
[2] INFN, Sezione di Catania, via Santa Sofia 64, 95123 Catania, Italy
[3] Physikalisches Institut, Nussallee 12, D-53115 Bonn
[4] Helmholtz-Institut für Strahlen- und Kernphysik, Nussallee 14-16, D-53115 Bonn
[5] INFN, Sezione di Pavia, via Agostino Bassi 6, 27100 Pavia, Italy
[6] Justus-Liebig-Universität Giessen, II. Physikalisches Institut, Heinrich-Buff-Ring 16, D 35392 Giessen
[7] National Science Center Kharkov Institute of Physics and Technology, Akademicheskaya St. 1, Kharkov, 61108, Ukraine
[8] INFN, Sezione di Roma "Tor Vergata", Via della Ricerca Scientifica 1, 00133 Roma, Italy
[9] Dipartimento di Fisica, Università di Roma "Tor Vergata", via della Ricerca Scientifica 1, 00133 Roma, Italy
[10] Dipartimento di Fisica, Università di Torino, via P. Giuria 1, 10125 Torino, Italy
[11] INFN, Sezione di Roma, c/o Dipartimento di Fisica - Università di Roma "La Sapienza", P.le Aldo Moro 2, 00185 Roma, Italy
[12] Istituto Superiore di Sanità, viale Regina Elena 299, 00161 Roma, Italy
[13] The University of Edinburgh, James Clerk Maxwell Building, Mayfield Road, Edinburgh EH9 3JZ UK
[14] Petersburg Nuclear Physics Institute, Gatchina, Leningrad District, 188300 Russia
[15] Russian Academy of Sciences Institute for Nuclear Research, prospekt 60-letiya Oktyabrya 7a, Moscow 117312 Russia
[16] Institut für Physik, Klingelbergstrasse 82, CH-4056 Basel

[17] INFN - LNF, Via E. Fermi 40, 00044 Frascati (Roma), Italy
[18] INFN, Sezione di Torino, via P. Giuria 1, 10125 Torino, Italy

E-mail: vdeleo@unime.it

Abstract. The results obtained at the BGO-OD experiment with the BGO calorimeter, equipped with the new electronic readout based on sampling ADCs, during the tests performed with the beam time of March and June 2012 are presented. The proper functioning of the apparatus has allowed the reconstruction of the pseudo-scalar mesons π^0 and η invariant masses. The simulation of the η' photoproduction reaction prepared for a proposal to the joint ELSA-MAMI Physics Advisory Committee is also presented.

1. Introduction

The excitation spectrum of the nucleon is the most important ground where to test our understanding of QCD in the strong-coupling regime and the study of baryon resonances plays the same role for understanding the nucleon structure as the nuclear spectroscopy played for the investigation of the nucleus structure.

There are major experimental, theoretical and computational efforts that aim to explore both the spectrum and the structure of excited nucleons. The dominant decay channel of nucleon resonances is the hadronic decay via meson emission [1]. Photoproduction of mesons, which carries information on strong and electromagnetic decay properties, therefore provides a very valuable tool for their study.

Dedicated experimental programs exist to perform accurate measurements of meson photo- and electro-production off the nucleon in order to discover its excitations, and determine its internal structure. A new generation of electron accelerators equipped with tagged photon facilities have opened the way to meson photoproduction experiments of high sensitivity and precision [2]. The BGO-OD experiment at the ELSA facility at Bonn (see Figure 1) involves the use of a Bremsstrahlung tagged and polarized photon beam of energy between 0.7 and 3.2 GeV, a large solid angle high resolution BGO calorimeter and the Open Dipole spectrometer equipped with tracking detectors. This apparatus will be used to measure polarisation observables and cross sections in the photoproduction of pseudo-scalar and vector mesons off an hydrogen or deuterium target.

2. Experimental Setup

A schematic view of the experimental apparatus located at the S-beamline of Bonn is shown in Figure 2. The Electron Stretcher Accelerator (see Figure 1) consists of three stages (injector LINAC, booster synchrotron and stretcher ring) and provides a beam of polarized and unpolarized electrons with a tunable energy up to 3.5 GeV. The extracted bunched electron beam impinges on a radiator in the S-beamline. Scattered electrons produce coherent and incoherent bremsstrahlung photons of energy $0.1 \div 3\,GeV$.

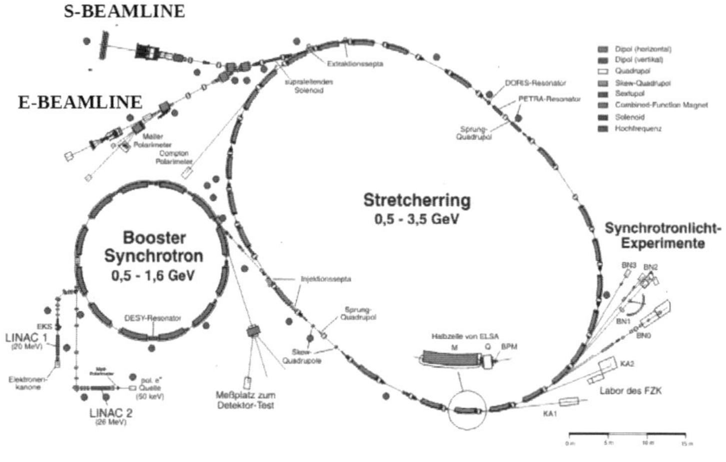

Figure 1. Elsa Facility.

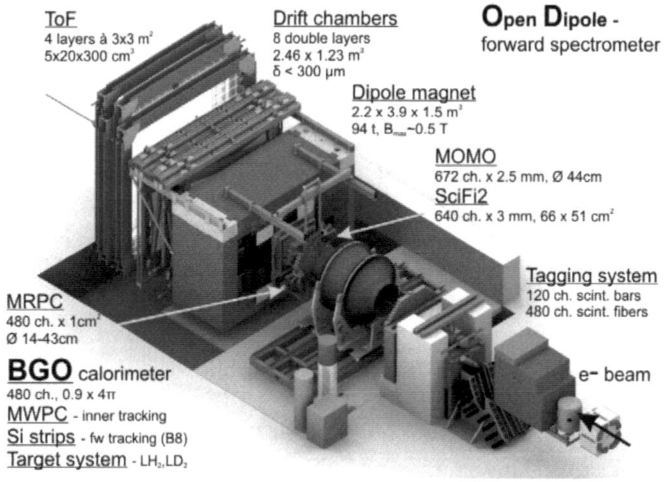

Figure 2. Schematic view of the Experimental Setup.

The BGO-OD detector setup is a combination of a central detector system and a forward spectrometer for charged particles, completed by a photon tagging system. The resolution of the tagger is about $1\% \div 2\%$ of the incident electron beam.

In the polar angular region between 25° and 155° we encounter:

- The BGO ($Bi_4Ge_3O_{12}$) Rugby Ball, a large acceptance calorimeter (see Figure 3) able to measure multi-photon event yield with excellent energy resolution (3% FWHM at 1 GeV). The design of the calorimeter has been performed in order to have a constant thickness in every direction and a central hole of radius $100\,mm$ for the passage of the beam, target and inner detector housing. The resulting structure is made of 480 truncated pyramidal crystals of 240 mm length (corresponding to 21 radiation lengths) arranged in a 15×32 matrix covering the polar angles from $25°$ to $155°$ and the whole azimuthal angles, corresponding to a total solid angle $\Delta\Omega$ of $11.3\,sr$. The mechanical structure consists of 24 carbon fiber baskets, each containing 20 crystals, and supported by an external steel frame. The baskets are divided into cells to keep the crystals mechanically and optically separated.

- A cylinder of 32 plastic scintillator bars, for the measurement of the Δ_E of charged particles, which allows, in combination with the energy released in the calorimeter, the discrimination between charged and neutral particles and the identification of charged particles (protons and pions).

- Two cylindrical multi wire proportional chambers (MWPC), under construction, for the reconstruction of trajectories from charged particles. Each chamber is made of two coaxial cylindrical cathodes which are segmented into strips, helically wounded in opposite directions. The anode array consists of equally spaced wires stretched parallel to the cylinder axis in the middle of the active area (gas gap $\sim 8\,mm$). The geometrical parameters are tuned to fit inside the carbon structure of the BGO and around the target nose.

Figure 3. BGO Calorimeter.

- The target cell and the cryogenic system, which are located directly behind the beam dump along the beam direction. The cell is contained in a vacuum pipe, which guarantees the thermal insulation and is surrounded hermetically by the central detectors. The target cell is a $4\,cm$ diameter Aluminum cylinder, closed by thin mylar windows at the two sides for the passage of the beam. Two different lengths of the cell are available (6 and $11\,cm$) in order to fulfill different experimental requirements. The target cell can be filled either with liquid Hydrogen (H_2) or Deuterium (D_2).

To cover the forward angular region between $\theta_{lab} = 25°$ and $\theta_{lab} = 8°$ an azimuthally-symmetric Multi-gap Resistive Plate Chamber detector (MRPC) will be installed between the BGO and the MOMO detector. The detector is divided into 16 phi sectors and it has in total 480 channels. Each sector in the azimuthal angle consists of two stacks. The expected spatial resolution σ_s is $1\,cm^2$ and the time resolution σ_t is about $50\,ps$.

The polar angular region of forward angles ($\theta < 12°$) is covered by the B1-magnetic field spectrometer that uses a dipolar field of about 0.5 T for the separation, identification and measurement of the momentum (3% resolution) of charged particles emitted in the photoproduction processes.

For this purpose, the spectrometer is equipped with:

- MOMO and SciFi2 that are scintillating fiber detectors for the reconstruction of the tracks in front of the magnet [6].

- Eight double layer drift chambers (DCs) which are built at PNPI of Gatchina, for the tracking of charged particles behind the spectrometer. These are arranged in four different orientations, the vertical wires to measure the x-coordinate, the horizontal ones to measure the y-coordinate, and tilted by $\pm 9°$ from vertical for u- and v-coordinate measurement.

- A time-of-flight (TOF) detector, which is an essential component for particle identification, because it provides time of flight measurements for charged particles. It covers the region $\theta < 8°$ and $\theta < 12°$ in the vertical and in the horizontal directions, respectively, at a distance of $5\,m$ downstream of the target [6].

3. BGO calibration procedure

The absolute calibration of the Bgo Rugby Ball was obtained using the 1.275 MeV photons (E_{source}) emitted by three ^{22}Na sources located inside the BGO cylindrical hole. The response of the 480 crystals was equalized using a procedure that sets all the photomultiplier (PM) gains, varying their high voltages. Thirty Wiener Sampling ADC modules have been used for the acquisition of the signals of the BGO detectors. The modules allow the extraction of the total deposited energy and the starting time of the signal. The equalization is strictly necessary, because a threshold on the hardware sum of the energy released in the BGO is used as a trigger for the experiment. After the equalization, for the same energy released, the response of the PM associated to each crystal will be located at the same ADC channel. For the calibration procedure an internal trigger which is provided by the ADC, when a signal overcomes a certain threshold, was used; it was chosen in order to guarantee the same statistics on all the crystals, which a global trigger on the hardware sum of the energies of all the crystals could not insure. In Figure 4 a schematic representation of the experimental chain of calibration is shown: the output signal from the phototube, coupled to the crystal, is sent to a mixer that delays, eventually attenuates and sums over the signals, and then it is sent to the ADC module for the readout.

To obtain the calibration constants for each channel, we fix the PM voltages so that the response to the energy of the second peak of the source is located at the channel 60 of the ADCs. The channel is fixed within a tolerance of $\pm 3\%$, which, in any case, does

Journal of Physics: Conference Series **424** (2013) 012008 doi:10.1088/1742-6596/424/1/012008

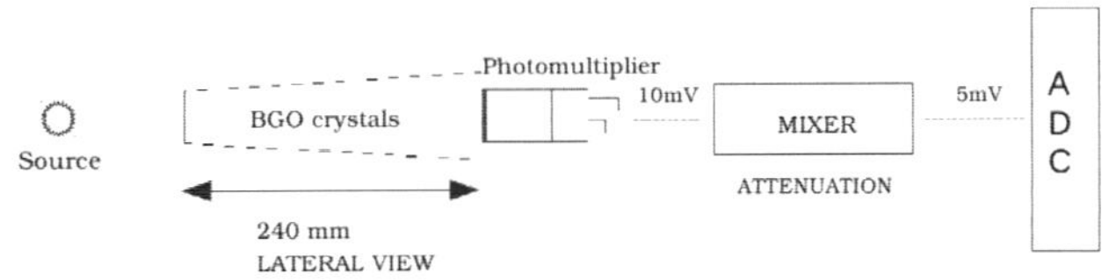

Figure 4. Scheme of the experimental calibration chain.

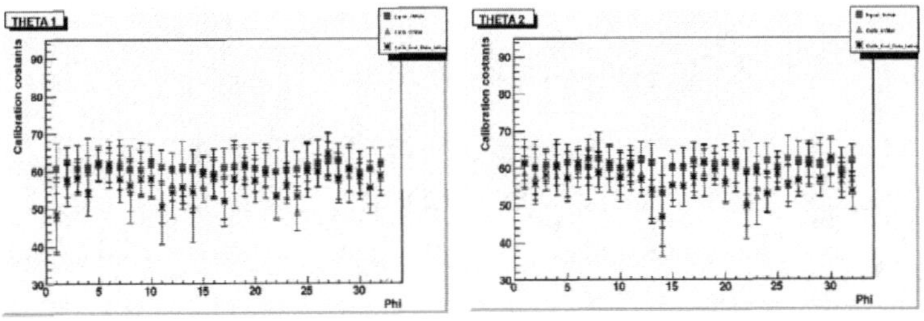

Figure 5. Calibration constants; the blue squares refer to the equalization on February 15, 2012 ; the red triangles to the calibration on March 1, 2012 ; the black stars to the calibration on March 5, 2012.

not affect the calibration of the data, since the calibration constant of each crystal is registered and used for the offline energy conversion.

The calibration procedure enables us to monitor the position of the peaks during the time (without changing the HV, just recording their position). It requires that the attenuators of the mixers and the beam are switched off.

The calibration constant of each crystal may change during time due to two principal reasons: variation in the crystal light output due to temperature effects and variation in the PM gain. Gain variations of the photomultipliers may occur as a consequence of the following effects:

-photocathode temperature variation;

-instability in the high voltage power supply;

-aging of the catode and dynode materials;

-voltage dividers instability [7].

In Figure 5 the azimuthal distribution ($\phi = 1 \div 32$) of the values of the calibration constants for two BGO crowns ($\theta = 1, 2$) related to different set of calibrations is shown. The blue squares refer to the equalization on February 15th, 2012; it is possible to see that all the responses are set within a maximum dispersion of $\pm 3\%$ (except for crystal $\theta = 2$, $\phi = 14$, whose PM had low gain). For the calibration constants on March 1st and March 5th, the fluctuations of the position of the second peak is about 1-2 channels which means an uncertainty of about 1.6%-3.2% of the energy. More fluctuations occur in the case of the comparison between the calibration constants obtained on February 15th and

March 1st. The position of the second peak changes of about six channels corresponding to an uncertainty of 9.8%. When a new calibration procedure is performed at a certain time and new calibration constants are recorded, these can be used for the correct conversion channels-energy at that time. But if strong differences with respect to the previous calibration occur, we can only infer that an uncertainty in the energy conversion arose in the period in-between. For this reason it is necessary to repeat the calibration procedure often in order to reduce the uncertainty in the behaviour of the calibration constants between two consecutive calibrations.

4. Preliminary results of the beam time tests.

The beam time tests were performed with the following experimental conditions: electron beam energy 2.5 GeV, two different BGO trigger conditions (the threshold of the energy sum of all the 15 crowns higher than 400 MeV and with a BGO-trigger where the threshold of the energy sum of the first 13 crowns was higher than 200 MeV), the target filled of liquid Hydrogen H_2, and the totally equipped BGO electronic readout. In Figure 6 a three-dimensional mapping of the BGO ball with the number of hits, when there was no beam and the accelerator was switched off, is represented. x and y coordinates are the number of the crown (θ) and the azimuthal distribution (ϕ) of crystals of the same crown, respectively. In the left part, the number of hits with energy threshold $> 1.0\,MeV$ is shown. Here, the three sources of ^{22}Na appear clearly. In the right part, the number of hits with energy threshold $> 20\,MeV$ is also shown.

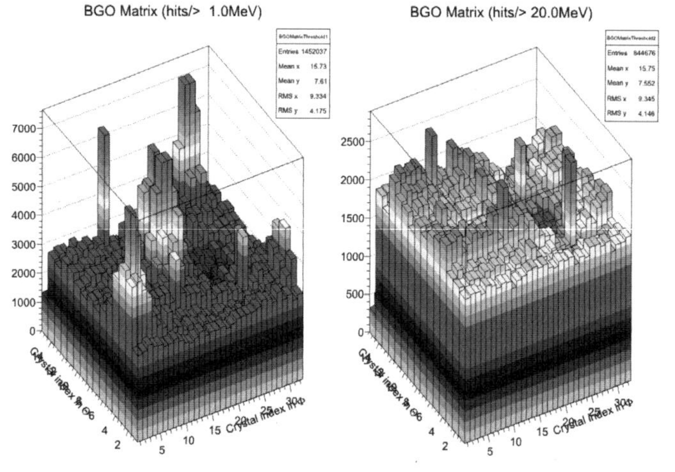

Figure 6. In the left part, the number of hits with energy threshold $> 1.0\,MeV$ is shown. Here, the three sources of ^{22}Na appear clearly. In the right part, the number of hits with energy threshold $> 20\,MeV$ is also shown.

In Figures 7 and 8 (indexes from $1 \div 240$ and from $241 \div 480$ correspond to the two halves of the BGO, left and right respectively, with respect to the incident beam) we show the energy and time distributions per index of crystals, respectively.

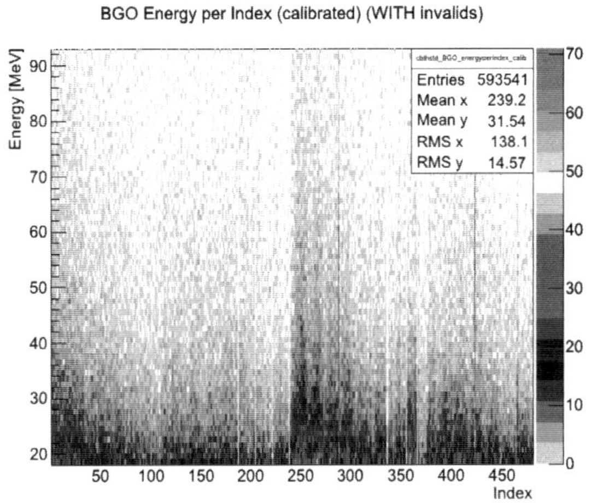

Figure 7. Energy distribution per index of the 480 crystals of the BGO.

As one can see, there is a strong global difference between the time measured in the two halves, which is trivially due to differences in the cable lengths, but also a little difference can be observed between the forward or backward crowns (indexes 1-48 and 192-240 for the left part; indexes 241-248 and 432-480 for the right part) probably which is probably due to the different electron transit time in the phototubes of different sizes. The analysis was performed with the Explora code [4] using the calibrated energy and time information of the deposited energy in the BGO crystals. To reconstruct the energy and position of incident particles in the BGO, the first step was the identification of photons from 'clusters' of adjacent crystals due to electromagnetic showers from bremsstrahlung and pair production.

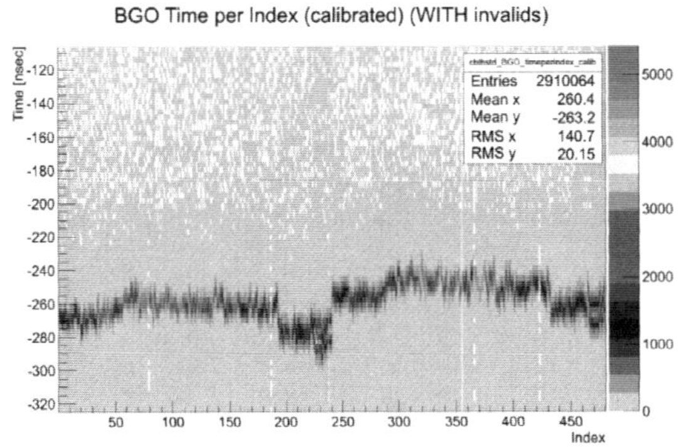

Figure 8. Starting time distribution versus the index of the 480 used crystals.

To identify the detected particles, without the charged particle identification (the scintillator barrel was still not in operation during the beam time), the size of reconstructed clusters within the BGO was used. A cluster of three or more crystals was assumed to be a candidate photon, a cluster with only one crystal is a possible candidate proton or a charged pion [5].

Particularly, we observed that when the total energy deposit is below about 300 MeV there is much low energy background to discern any structure in the spectrum. However above approximately 400 MeV there is a concentration of events with the invariant mass close to the expected π^0 mass, and a smaller shoulder above 500 MeV is consistent with the mass of the η meson.

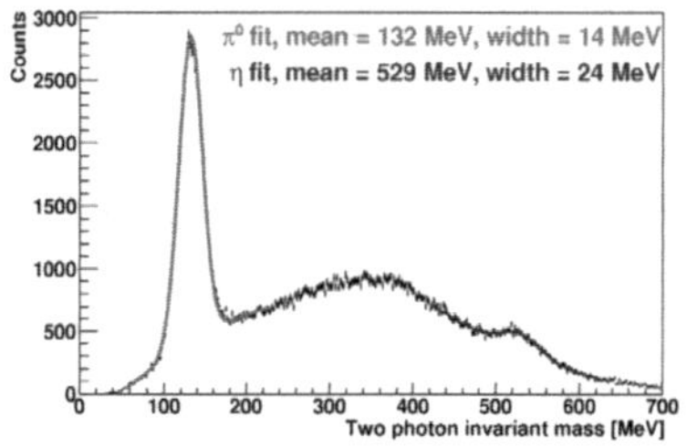

Figure 9. Two photon invariant mass spectrum taken with the BGO calorimeter in March 2012. [5, 6]

In Figure 9, the invariant mass of the two candidate photons is plotted for a total deposited energy $> 400\,MeV$.

5. Monte Carlo estimation of the Σ beam asymmetry in the $\gamma + p \to \eta' + p$ experiment planned at BGO-OD

Polarization observables in pseudo-scalar meson photoproduction have proved to be very efficient to fix parameters of resonances involved in the process [1, 2, 8, 9].

In the case of η' photoproduction off the proton, only total and differential cross section values are available in the literature (CLAS experiment at JLab [10, 11], and CB-ELSA-TAPS experiment at ELSA-Bonn [12]).

Two theoretical approaches were developed to describe these data [8, 9]:
i) in a relativistic meson-exchange model of hadronic interactions [8], Nakayama and Haberzettl include mesonic t-channel (ρ and ω) together with nucleon s- and u-channel and resonances contributions. The included resonances were $S_{11}(1535)$, $P_{11}(1710)$, $D_{13}(1520)$ and $P_{13}(1720)$, the two latter being required to reproduce some details of the angular distribution;
ii) in a reggeized model for η and η' photoproduction [9], the above mentioned Authors used essentially the same ingredients but vector meson exchanges are treated in terms of Regge trajectories to comply with the correct high-energy behaviour.

Both approaches give a reasonable description of data and in both cases the authors stressed that the cross section data alone are insufficient to pin down the resonances

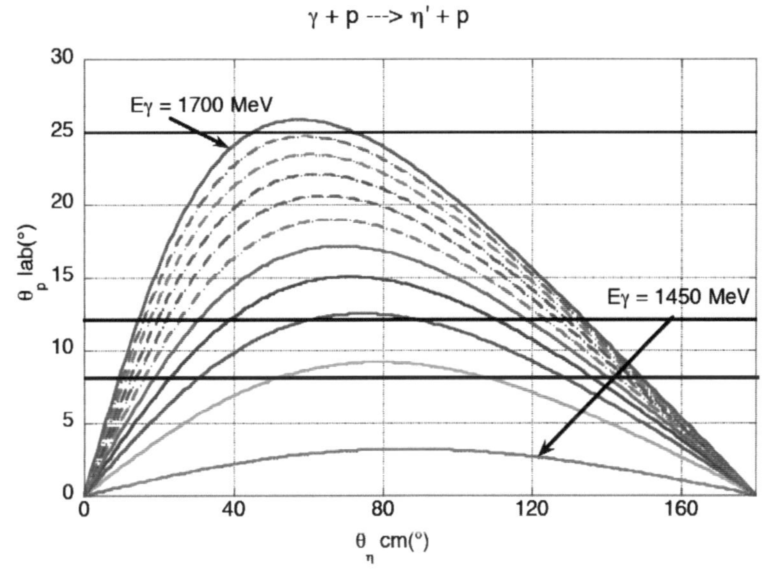

Figure 10. Kinematics of the reaction $\gamma + p \rightarrow \eta' + p$. The curves show (25 MeV steps for the incident photon energy E) the behaviour of the proton laboratory angle θ_p as a function of the $\theta_{\eta'}$ center-of-mass angle.

parameters, while beam and/or target asymmetries could be very helpful to better determine the partial wave contributions in this reaction, and impose more stringent constraints on the parameter values.

BGO-OD shows an excellent detection efficiency for both neutral and charged final states and will provide a polarized photon beam, allowing for the determination of still unmeasured Σ beam asymmetry in many reaction channels.

A very simple trigger is required to perform this measurement: the coincidence between a signal in the tagger and a constraint in the total energy collected by the BGO calorimeter, the latter is required to be greater than a threshold value $E_{BGO-thr}$ =180-200 MeV. This simple trigger scheme proved to be very efficient in reducing the electromagnetic background (mainly pair production on the target cell and along the photon beam line) without affecting the hadronic count rate. In Figure 10 the kinematics of the reaction is displayed. We can see in this figure that, in our required conditions, the recoil proton will always be measured either in the dipole spectrometer, or in the MRPC position chambers. A complete measurement is thus feasible by detecting the recoil proton in the forward direction ($\theta_p < 25°$) and the η' decay products in the BGO calorimeter. We will detect the η' via the following decay chains:

1. $\eta' \rightarrow \gamma\gamma$;
2. $\eta' \rightarrow \eta\pi^0\pi^0 \rightarrow 6\gamma$;
3. $\eta' \rightarrow \eta\pi^+\pi^- \rightarrow 2\gamma\pi^+\pi^-$.

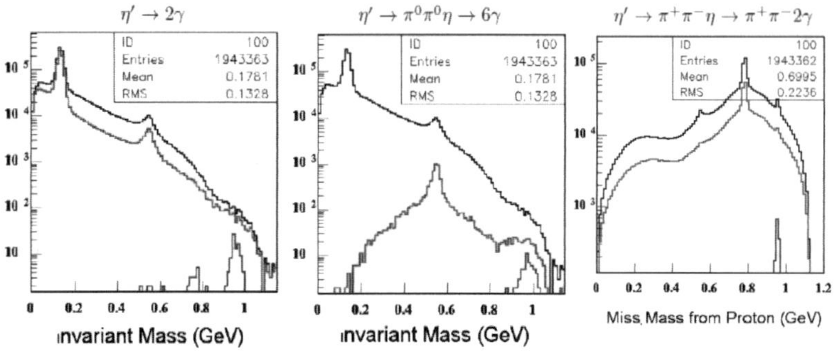

Figure 11. Event selection: the black curve is identical in left and central graphs and indicates the invariant mass spectrum of all events containing at least two photons measured in the BGO calorimeter, while in the right panel it indicates the missing mass from the detected forward proton. The red curve shows the same spectrum when exactly 2 photons(left), six photons(central), 2 photons and 2 identified charged pions (right) are reconstructed in the calorimeter. The blue curve shows the remaining events after the kinematical constraints from the recoil proton measurement are applied. In all cases (neutral and charged channels) the η peak is cleanly selected.

For each decay chain the detection efficiency must be estimated. We have used a full MonteCarlo (MC) simulation of the detector setup that includes an event generator (updated version of the one described in [13]) containing all known hadronic cross sections off the proton as a function of the photon energy.

The result of the simulation and event selection procedure is summarized in Figure 11. In this figure we can see the yields (in arbitrary units) at various stages of the selection procedure: the black curve in all graphs represents the starting point of the event selection when at least 2 photons are detected in the BGO. In the neutral decay channel the main variable to identify the η' events is the invariant mass reconstructed from the detected γ's in BGO (left and central graphs in Figure 10), while in the charged decay the main variable is the missing mass to the detected and identified forward proton, because our apparatus is not able to directly measure the energy of charged pions in all solid angle (right panel in Figure 10). The red curve indicates the resulting spectrum when exactly 2γ (left), 6γ (central), 2γ and 2 identified charged pions (right) are measured, and finally the blue curve shows the events that fulfil the kinematical constraints in the recoil proton measurement. The values of the reconstruction efficiencies depend on the percentage of background events that are accepted by the event selection procedure. By simulation, we estimate in accurate way that the analysis efficiency (taking into account the competition of all concurrent channels [13]) is about 25% for 2 γ channel, 4.4% for $\eta\pi^0\pi^0 \to 6\gamma$, and 5.4% for $\eta\pi^+\pi^- \to 2\gamma\pi^+\pi^-$.

With the aim to measure the beam asymmetry from threshold up to 1.7 GeV with an error $\Delta\Sigma$ of 0.07 in five energy and six angular bins, we were able to estimate that the experiment is feasible with the reasonable 1000 hours beam time, by using the 6 cm length liquid Hydrogen target and a photon beam linearly polarized through coherent bremsstrahlung with a total tagged intensity of N = 5 $\times 10^7 s^{-1}$.

6. Conclusions

During the February/March beam time 2012, for the first time, tests on totally equalized and calibrated BGO detector were performed with a full equipped BGO electronics consisting of 30 ADCs AVM-16 Mambo from Wiener. The calibrated energy and time of energy deposits in the BGO crystals have been used in the analysis. No charged particle identification was available (no data from the plastic scintillator barrel or MWPC) and no data from the photon tagger or any information from the forward spectrometer was used in the analysis. The reconstruction of the pseudoscalar mesons η and π^0 from the decay photons was achieved. It was done plotting the two photon cluster invariant mass spectrum for events with at least 400 MeV energy deposit yields. It was performed an accurate feasibility study on the $\gamma + p \rightarrow \eta' + p$ reaction, demonstrating the possibility to perform a Σ beam asymmetry measurement at BGO-OD experiment in the energy range starting from the threshold up to 1.7 GeV. We expect that the measure of this observable will be useful to contribute to solve the puzzle related to right weights of different resonances [8, 9] involved in the reaction process, unsolved by the knowledge of the differential cross section [10–12].

References

[1] Mandaglio G *et al.* 2010 *Phys.Rev C* **82** 045209.
[2] Krusche B *et al.* 1999 *Czech. J. Phys* **49**.
[3] http://bgo-od.physik.uni-bonn.de/ExperimentalSetup.
[4] Hartmann J, Piontek D *et al.* 2011 *Explora code:version 1.3*.
[5] Jude T, De Leo V, Curciarello F 2012 *Internal Note* - BGO-OD Collaboration.
[6] http://bgo-od.physik.uni-bonn.de/ExperimentalSetup/BgoBall.
[7] Ghio F, *et al.* 1998 *Nucl. Inst. and Meth. A* **404** 71-86.
[8] Nakayama K, Haberzettl H 2006 *Phys. Rev C* **73**, 045211.
[9] Wen-Tai Chiang *et al.* 2003 *Phys. Rev C* **68** 045202.
[10] Dugger M *et al.* 2006 *Phys. Rev. Lett.* **96** 062001.
[11] Williams M *et al.* 2009 *Phys. Rev. C* **80** 045213.
[12] Crede V *et al.* 2009 *Phys. Rev. C* **80** 055202.
[13] Corvisiero P 1994 *et al. Nucl. Inst. and Meth. A* **346**,433.

Meson photoproduction and baryon resonances at MAMBO experiment

Mariia Romaniuk[1,2]

[1]Dipartimento di Fisica e di Scienze della Terra, Universita di Messina, I-98166, Messina, Italy
[2]INFN- Sezione Catania, I-95123, Catania, Italy

E-mail: mromaniuk@unime.it

Abstract. Photoproduction of mesons within the framework of the MAMBO experiment (BGO-OD at Bonn plus MAMI at Mainz) was studied. The results on the operative work of the cryogenic H_2/D_2 target system during the last commissioning beam times at the March and June 2012 are shown. Investigation of the single charged pion photoproduction was provided using a polarized ^{3}He target at the tagged photon facility of the MAMI accelerator. Unpolarized and helicity dependent cross sections are presented for channels $\gamma N \rightarrow \pi^{\pm} X$ in the $\Delta(1232)$ baryon resonance region.

1. Introduction

During last 40 years Quantum Chromo Dynamics (QCD) established as the theory of the strong interaction. Perturbative approach could be used in the high energy regime, where the strong coupling constant is small, and in fact this approach has been successfully tested in several experiments. But in low energy regime, which is typical for nucleon and its excited states, it is not possible to use such approach because the strong coupling constant becomes large.

From the number of the effective degrees-of-freedom and their quantum numbers follow the number of the excited states with definite quantum numbers. The dominant decay mode of the nucleon resonances is hadronic decay via emission of mesons with a lifetimes typical for strong interactions $\tau \approx 10^{-24}s$ and a corresponding widths of a few 100 MeV [1]. Such a width together with spacing of the resonances not more than 10 MeV makes it difficult to investigate individual states, see Fig. 1.

For detailed tests of the quark models it is not enough studying of the excitation spectra (excitation energies and quantum numbers of the states). More information could be extracted from the transitions between the states which reflect their internal structure and are more sensitive to the model wave functions. Photoproduction of mesons is particularly interesting because the rich information connected to the electromagnetic transition amplitudes can be accessed in addition to the dominant hadronic decay modes [2].

In the present work the experimental test of nucleon models were performed in a two way. From one hand, the problem of the missing resonances could be studying by a large scale survey investigating many different final states – $N\pi$, $N\pi\pi$, $N\eta$, $N\eta'$, $N\omega$, $N\rho$ etc – over a large energy scale (BGO–OD). On the other hand, the low lying resonances can be studied in great detail providing data for precision tests of the models. The availability of linearly and circularly

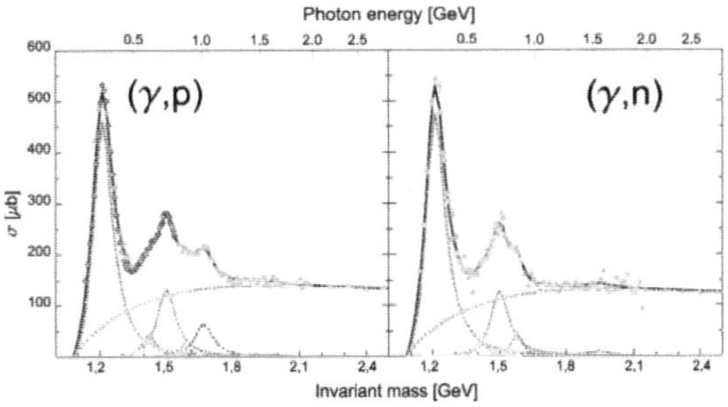

Figure 1. Total photoabsorption cross section for reaction $\gamma N \to NX$; left side – on proton and right side – on neutron. Points are measured data, curves – fit of Breit-Wigner shapes of nucleon resonances $P_{33}(1232)$, $P_{11}(1440)$, $D_{13}(1520)$, $S_{11}(1535)$, $F_{15}(1680)$ (only for proton) and $F_{37}(1950)$ and a smoothly varying background [1].

polarized photon beams together with polarized targets has provided access to observables, which are sensitive to specific resonances (MAMI).

2. BGO–OD

The new experimental setup of the recently established BGO–OD collaboration consists of the combination of an open-dipole forward spectrometer and the BGO ball which cover the central angular region. This configuration is ideally suited to investigate the photoproduction of multi-particle final states with mixed charges. In addition it will allow nucleon polarization measurements in single-meson photoproduction. Due to the excellent forward acceptance it opens the possibility to investigate vector-meson production in order to understand the reaction mechanism and the role of resonances. The BGO–OD collaboration presently includes individuals and groups from Germany (Bonn), Italy (Rome, Frascati, Pavia, Messina), Russia (Gatchina, Moscow), UK (Edinburgh, Glasgow) and Ukraine (Kharkov). The experimental set-up (see Fig. 2) consists of a large 90 ton dipole magnet, tracing detectors, two scintillating fiber detectors, MOMO and SciFi2 (to allow for momentum reconstruction of charged particles bent through the magnetic field), an aerogel Cherenkov detector (discriminates pions against protons and particularly improves the K-identification substantially), a time-of-flight (TOF) detector (provides flight-time measurements for charged particles and neutrons), the BGO Ball which is optimized for photon detection, good performances with protons (energy and angle reconstruction) and good neutron detection efficiency. BGO hermetically encloses the target (polar angular range 25–155 degrees). Main features of BGO–OD beam: beam energy $E_\gamma = 0.7 - 2.8$ GeV; linear polarization degree up to $2.0 - 2.2$ GeV; high intensity ($10^7 \gamma/s$) for low cross section reactions [3].

The Quantumcooler Hydrogen/Deuterium is a two stage closed-cycle refrigerator. The working fluid in all stages is an high purity helium gas. The nominal operating temperatures of each stage is 70 K and 20 K, respectively. Both stages are based on the Gifford-McMahon (G–M) cycle. The target cell is a 3 cm diameter aluminum cylinder, closed by thin mylar windows at the two sides. Two different lengths of the cell 6 cm and 11 cm are available. The target cell could be filled with liquid Hydrogen (H_2) or Deuterium (D_2). The hydrogen/deuterium gas is cooling down by the helium through heat exchangers and liquefied inside the cell.

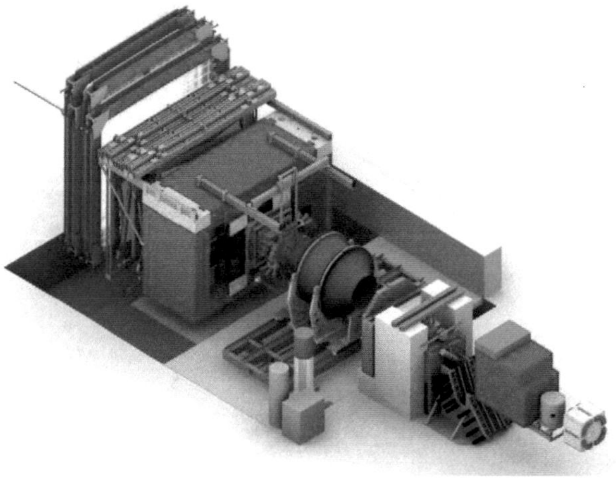

Figure 2. Overview of the experimental set up of the BGO–OD experiment

Starting from room temperature the target started to fill after 5 hours and full normally after 8 hours. To empty/fill cell is possible not only by "hand" changing the temperature of heater from 18 to 25 K, but also by remote, using the "slowcontrol" software. This process can be conveniently monitored by observing the H_2 tank pressure which varies from +0.9 *bar* (1.9 *bar* absolute) with the target empty to +0.74 *bar* (1.74 *bar* absolute) with the target full, see Fig.3.

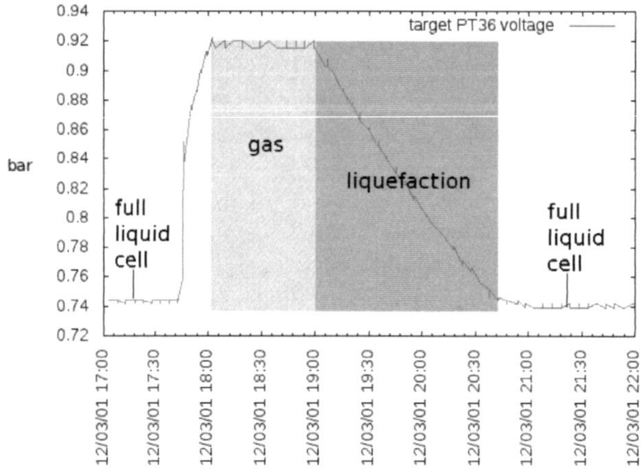

Figure 3. Pressure on the gas tank during empty/refill process, 0.9 *bar* corresponds to the empty cell, 0.74 *bar* — to the full with liquid Hydrogen cell.

This has allowed in March and June 2012 to perform preliminary data taking for the commissioning of the BGO calorimeter, studying the π^0 and η photoproduction on the proton and comparing the results for full/empty target. First results on the two–photon invariant mass have shown the detection of the π^0 and η decay inside the BGO calorimeter. Further refinements are required on the beam collimation which still shows a strong background contamination.

3. MAMI

Another important subject in study of the baryon resonances is helicity dependent variables (especially on neutron), which provide better understanding of the reaction mechanism and structure effects. ^3He is a system consisting of two protons with spins paired off and an active unpaired neutron, in relative s state with 90% probability. As a result,magnetic moment of ^3He can be approximated as the magnetic moment of the neutron: $\mu_{^3He} \approx \mu_n$.

During July, 2009 a first measurement of the double polarized photoabsorption cross section on ^3He in the $\Delta(1232)$ baryon resonance region has been performed at the tagged photon facility of the MAMI accelerator in Mainz. Circularly polarized photons were obtained by bremsstrahlung of longitudinally polarized electrons with an average polarization of 75%, the bremsstrahlung photons were tagged using the Glasgow-Mainz magnetic spectrometer (energy resolution $\sim$ 2 MeV). The relative tagging efficiency was monitored by a CCD photon camera and absolute measurements were made using a lead-glass detector. The reaction products were detected by the central detector system (see Fig. 4) consisting of the Crystal Ball (CB) NaI spectrometer, complemented by the Multi-Wire Proportional Chambers (MWPCs), used to identify and track the charged particles, and the cylindrical Particle Identification Detector (PID), used to distinguish the charged from the neutral particles detected by the CB. The combined information provided by these three detectors provides accurate energy, angle and particle identification in the azimuthal (φ) and polar (θ) angular regions from 0° to 360° and from 21° to 159°, respectively.

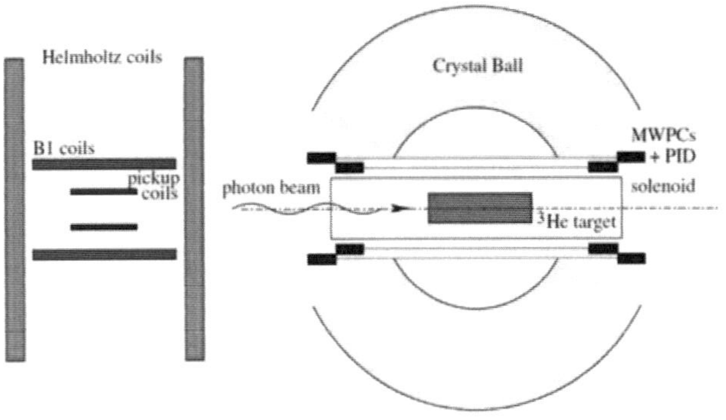

Figure 4. Side view of the central detectors setup [4]

Results and conclusions

Using dE/dx vs E technique, PID and MWPCs information pion/proton separation has been done. Preliminary results for the unpolarised and polarised differential cross sections of the $\gamma\,^3He \rightarrow \pi^{\pm}X$ reactions are presented in Fig. 5, 6. The data are compared to the predictions from Fix–Arenhoevel model [5] that take into account nuclear effects, such as FSI, and to the MAID model, where the nucleons inside the nucleus are assumed to be free and only the effect of the spin alignments are taken into account. The polarised data are also compared to the DAPHNE data obtained on deuteron [6]. There is a good agreement for polarised differential cross sections between ^3He and deuteron data. The nuclear structure contributions are less important than for unpolarised case.

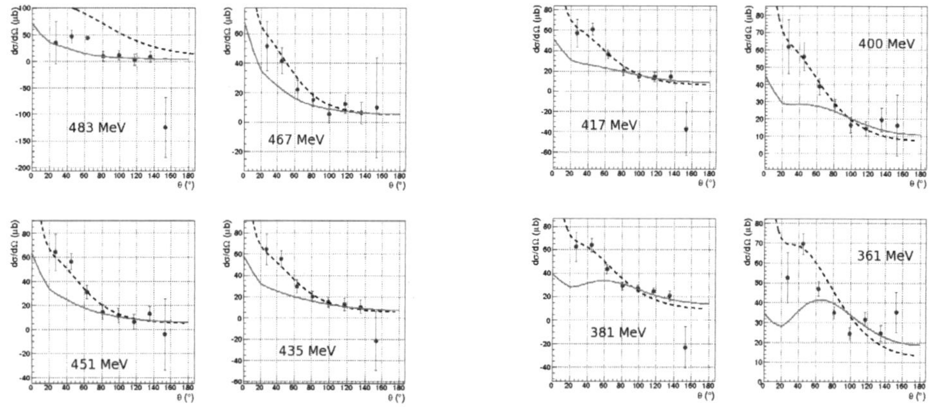

Figure 5. Preliminary results for the differential unpolarised cross section for the semi–inclusive channels $\gamma\,^3He \rightarrow \pi^{\pm}X$ for $E_\gamma = 361 - 483$ MeV. The experimental data (blue circles) are compared to the MAID model (dashed line) and Fix-Arenhoevel model (continuous red line).

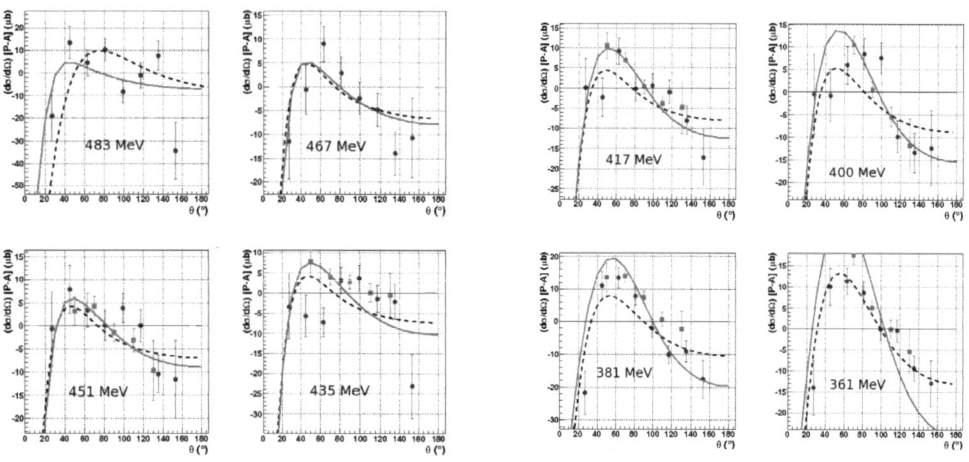

Figure 6. Preliminary results for the helicity dependent differential cross section $\Delta\sigma$ for the semi–inclusive channels $\gamma\,^3He \rightarrow \pi^{\pm}X$ for $E_\gamma = 361 - 483$ MeV. The experimental data (blue circles) are compared to the MAID model (dashed line), Fix-Arenhoevel model (continuous red line) and DAPNE data [6] (red squares).

The present theoretical models are not able to describe in a satisfactory manner the experimental results. It is need to perform better description of the considered processes. Further measurements to improve statistics are needed.

References

[1] Bianchi N et al. 1996 *Phys. Rev. C* **54** 1688
[2] Krusche B, Schadmand S 2006 *arXiv:nucl-ex/0306023* 1
[3] ELSA: *Letter to intent to the PAC* 2009.
[4] Pedroni P 2011 *Eur. Phys. J. Topics* **198** 181
[5] Fix A *Private communication* 2012
[6] Ahrens J et al. 2010 *Eur. Phys. J. A* **44** 189